Complex Fluid-Flows in Microfluidics

Francisco José Galindo-Rosales

Editor

Complex Fluid-Flows in Microfluidics

Editor
Francisco José Galindo-Rosales
Centro de Estudos de Fenómenos de
    Transporte
Faculdade de Engenharia da Universidade
    do Porto
Porto
Portugal

ISBN 978-3-319-86658-1   ISBN 978-3-319-59593-1   (eBook)
DOI 10.1007/978-3-319-59593-1

Printed on acid-free paper

This Springer imprint is published by Springer Nature
The registered company is Springer International Publishing AG
The registered company address is: Gewerbestrasse 11, 6330 Cham, Switzerland

To my three years old son (Diego),
the best mentor I ever met.

Francisco José Galindo-Rosales

Foreword

This book is a result of the efforts put by the authors in the preparation of the First Summer School on Complex Fluid-Flows in Microfluidics (10–14 July, 2017, University of Porto, Portugal). This five-day course includes five topics, which are summarized in the main chapters of this book:

1. Complex fluids and rheometry in microfluidics;
2. Microfabrication techniques for microfluidic devices;
3. Fluid-flow characterization in microfluidics;
4. Numerical simulations of complex fluid-flows at microscale;
5. Numerical optimization in microfluidics.

This is a concise and timely book, on an exciting area of research that has seen a tremendous increase of interest with the advent of cheap and reliable microfabrication techniques. The small length scales, typical of microfluidics, enhance the nonlinear response of complex fluid-flows, and this characteristic can be explored in several applications, from micro-rheometry techniques, to the investigation of purely elastic instabilities or elastic turbulence in low Reynolds number flows. This book provides a description of experimental and numerical techniques for complex fluid-flow analysis and is particularly useful as an introduction to the field of microfluidics and to the basics of rheology and complex fluid-flows. Some advanced topics are also covered, such as the numerical optimization of microfluidic devices for complex fluid-flows. This is a topic with a wide range of practical applications, but so far has only been explored in a few relevant applications, such as the development of efficient micro-rheometers or efficient devices for manipulation of biological samples or complex fluids.

April 2017
Manuel A. Alves
University of Porto, Porto, Portugal

Preface

This book is written to serve as textbook for the *1st Summer School on Complex Fluid-Flows in Microfluidics* held in the Faculty of Engineering of the University of Porto (Portugal, 2017). Moreover, we also expect it to be useful as a pedagogical introduction for any researcher starting to work on microfluidics in combination with complex fluids. Having this latter and broader objective in mind, this book covers the fundamentals of the theoretical, experimental, and numerical approaches distributed in five chapters: The first one is fully dedicated to complex fluids and the role that microfluidic can play as a platform to perform rheological characterization beyond the limits of the macroscopic rheometers; the second one focus its attention to the different experimental techniques to develop microfluidic devices; the third ones details and assesses the advantages and disadvantages of the main the experimental techniques for the characterization of the complex fluid-flow at microscale; the forth and the fifth ones are fully oriented to numerical aspect, that is, computational simulations of the fluid-flow and numerical optimization, respectively. In every chapter, the reader will find ten practical advices that any novice should follow to avoid problems when dealing with complex fluid-flows at microscale.

I am personally grateful to all the co-authors for their commitment and generous effort in preparing their contributions to this book.

This book has been edited/written during the FCT Investigator Grant of the Fundação para a Ciência e a Tecnologia (IF/00190/2013).

Porto, Portugal
April 2017

Francisco José Galindo-Rosales

Contents

Chapter 1
Complex Fluids and Rheometry in Microfluidics

Francisco J. Galindo-Rosales

Abstract Complex fluids are everywhere, literally, just need to look around you, or even closer, inside your own body. These fluids are named *complex* because when they flow, they do not hold a linear relationship between the rate of deformation and the stress tensors, and consequently the Newton's law of viscosity is not suitable for them. In this chapter, the importance of the performing a rheological characterization and choosing the right constitutive model is highlighted, in particular when flowing at microscale, where the elastic behavior of these complex fluids is enhanced even at very small Reynolds numbers. Additionally, the potential of microfluidics as a platform for performing rheological characterizations is tackled.

1.1 Introduction

If we could depict a spectrum of mechanical behaviours of materials (Fig. 1.1), on the left side we would find Elastic Solid Materials as such exhibiting a linear relationship between the deformation tensor and the stress tensor, i.e. following Hooke's law of elasticity; on the right side of that spectrum we would have the Newtonian Fluids, as those materials that exhibit a linear relationship between the rate of deformation tensor and the stress tensor under flow, i.e. they follow Newton's law of viscosity. Any other material with the ability to flow would be placed in between these two extreme boundaries and is susceptible to be considered a *complex fluid*. Rheology is a branch of science dedicated to the study of deformation and flow of matter, it deals with materials that are neither Hookenian nor Newtonian, meaning that complex fluids in general range from *soft solids* to *elastic liquids* [8, 51, 56].

Complex fluids can be classified either as field-passive or field-active materials, depending on whether they need of an external field (magnetic or electric) to trigger their rheological behaviour or if their non-linear flow behaviour is simply shown upon the application of an external load, respectively [38]. In this chapter, the attention

F.J. Galindo-Rosales (✉)
Centro de Estudos de Fenómenos de Transporte, Faculdade de Engenharia da Universidade do Porto, Rua Dr. Roberto Frias, S/n, CP 4220-465 Porto, Portugal
e-mail: galindo@fe.up.pt

© Springer International Publishing AG 2018
F.J. Galindo-Rosales (ed.), *Complex Fluid-Flows in Microfluidics*,
DOI 10.1007/978-3-319-59593-1_1

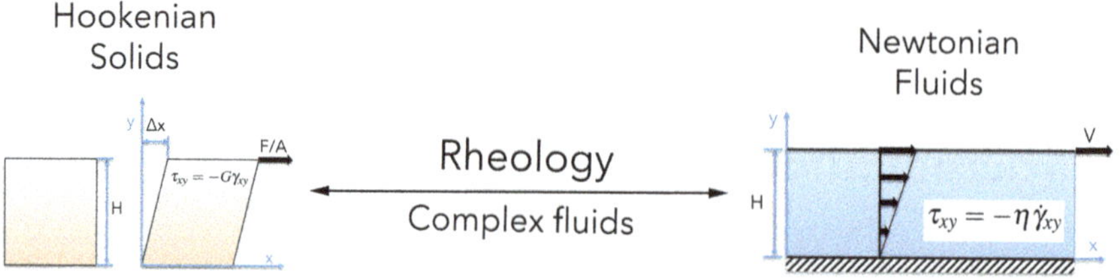

Fig. 1.1 Spectrum of mechanical behaviours of materials, ranging from Hookenian solids to Newtonian fluids. Rheology connects these two ideal mechanical behaviours and focus on the study of the deformation and flow of complex fluids, from soft solids to elastic liquids

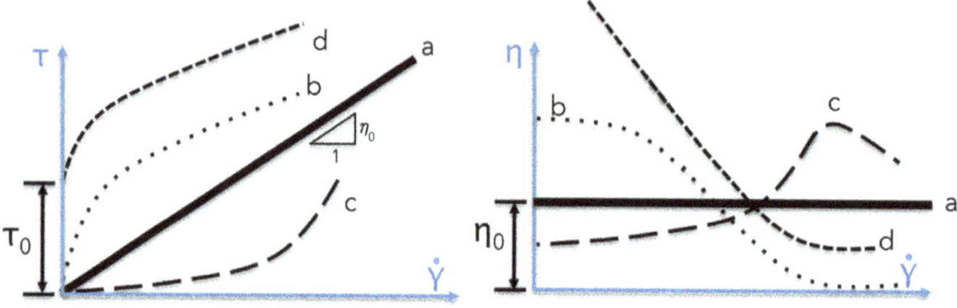

Fig. 1.2 Flow curve (*left*) and viscosity curve (*right*) of a Newtonian fluid **a** and three Generalized Newtonian Fluids: **b** Shear thinning fluid, **c** Shear thickening fluid, and **d** Yield Stress fluid

will be focused on field-passive materials, but readers interested in the rheological behaviour of field-active materials and their applications in microfluidics can find more information in the reviews of Sheng [67] and Nguyen [58]. A classical example of field-passive complex fluid is the silly putty, which exhibits solid-like behavior when subjected to quick mechanical deformations, while it flows like a thick liquid when the time of deformation is rather large. Greek yogurt is another traditional example of a field-passive complex fluids behaving like a soft-solid at rest while flowing like a liquid under sufficiently large shear stresses.

Typically, the source of the non-linearity between stress and rate of deformation tensors in a complex fluid has to be found in its formulation, that is either in the presence of molecular chains or in the existence of aggregates of nanoparticles. The way in which they stretch, orient and even break due to the flow dynamics and how they relax upon the release of the stresses is responsible for the bulk rheological behaviour of the fluid. The rheological behavior serve to sort complex fluids into two major groups: Generalized Newtonian Fluids (GNF's) and Viscoelastic Fluids (VEFs). GNF's show a non-linear relationship between the shear stresses and the applied shear rates, given by their apparent viscosity, which is not a constant, but a function of the shear rate. The dependency of the apparent viscosity with the shear rate allows the definition of different rheological behaviours, i.e. shear thinning, shear thickening and yield stress (Fig. 1.2).

GNF's are always be considered as inelastic fluids. Only VEF's exhibit simultaneously properties of elastic solids and viscous liquids, and therefore a characteristic relaxation time (λ) [54].

That characteristic relaxation time is responsible for the definition of the two most representative dimensionless numbers in rheology [60]: the Deborah number, defined as the relationship between the relaxation time and the duration of the deformation ($De = \frac{\lambda}{T}$); and the Weissenberg number, defined as the relaxation time multiplied by the rate of deformation ($Wi = \frac{\lambda U}{L}$). While the Deborah number allows distinguishing between solid-like ($De \uparrow$) and liquid-like ($De \downarrow$)behaviors of a particular material experiencing a deformation over a given timeframe, the Weissenberg number represents the ratio of elastic to viscous forces when a complex fluids is under flow. In general, Deborah and the Weissenberg numbers are not equivalent and the Weissenberg number is the one representing the nonlinearity of the rheological response of a viscoelastic material [22]. Thus, the steady, incompressible isothermal flow of a viscoelastic fluid is fully defined by the Weissenberg number and the Reynolds number ($Re = \frac{\rho U L}{\eta}$), being the latter the ratio of the inertial to viscous forces. The ratio between them defines the Elasticity number ($El = \frac{Wi}{Re} = \frac{\lambda \eta}{\rho L^2}$), which represents the ratio of elastic to inertial forces in the flow of a viscoelastic fluid [70].

Even for fluids with small relaxation times, microfluidics allows the exploration of zones in the $Wi\text{-}Re$ parameter space unreachable at macroscale, due to the very small characteristic length scale that enhances the elastic effects ($L \downarrow \Rightarrow El \uparrow$) [65].

Additionally, for the same reason, elastic instabilities may be triggered with relative ease, an consequently the flow of viscoelastic fluids can be very different from their Newtonian and Generalized Newtonian counterparts [33]. This intrinsic capacity for enhancing the elastic behavior of the viscoelastic fluids at low Reynolds numbers makes of microfluidics an unrivalled platform for both (1) performing rheological characterization beyond the limits of the commercial rheometers at macroscale [32, 69], and (2) for designing microfluidic rectifiers, i.e. microchannels with anisotropic flow resistance so that they can operate as fluidic devices (flux stabilizer or bistable flip-flop memory) similar to their solid-state electronic counterparts [40, 41] or they can be used for optimizing the mechanical performance of reinforced composites [36, 37]. Additionally, elastic instabilities can also be exploited for mixing enhancement at low Reynolds numbers [39] and enhanced oil recovery applications [14, 31]. Moreover, microfluidics allows for isolating individual polymer molecules in precisely defined flows and assessing models for polymer dynamics [81].

In this chapter, the attention will be focus on the potential of microfluidics as a platform for performing rheometry of complex fluids.

1.2 Rheometry at Macroscale

Within the world of Rheology, the field of Rheometry deals with the experimental determination of the rheological properties of complex fluids. It is necessary to perform experimental tests under controlled deformations/stresses and flows and record the mechanical response of the sample. As it is done in other disciplines, such as solid mechanics, these experiments must be performed according to some standards in order to measure some material functions, like the viscosity, and allow sharing and comparing that information either for quality control, for quantitative analysis or modeling [56]. Thus, rheometers are commercially available devices allowing to perform rheometry at macroscale imposing a small set of standard flows; namely, simple shear and extensional flows. Figure 1.3 summarizes the most common techniques to fully characterize rheologically complex fluids at macroscale, from low viscoelastic liquids to polymer melts. The reader interested in learning more about standard rheometry can get deeper through specialized books [10, 50, 51, 55, 56, 77].

Despite the correct use of the commercially available rheometers would allow getting a full rheological characterization of most of the complex fluids, in some cases it can be very challenging and we may find some limitations. Recently, R.H. Ewoldt and co-workers [26] thoroughly revised and summarized all these challenges and limitations associated to the shear rheometry at macroscale. Some of these limitations can be overcome by scaling down to microscale:

1. Instrument specifications. As any experimental device, rheometers are limited by their intrinsic measurable ranges of load and displacement. Many viscoelastic fluids exhibit relaxation times rather small (≤ 1 ms) and, consequently, their viscoelastic stresses are small in laminar macroscale flows. Thus, it becomes very difficult to measure their material functions in a conventional rheometer ($El \ll 1$). As mentioned above, working at microscales may help on solving this limitation, due to the enhancement of the elastic effects ($El >> 1$) [69].

2. Volume sample. In many cases, mainly when dealing with biological fluids, the availability of samples may be limited to very small quantities. Working with rheometrical techniques in which the characteristic length scale is well below 1 mm would allow to reduce extremely the consumption of volume samples to perform the experiments (≤ 1 μl) [46, 47], without detriment to the accuracy in the measurements.

3. Instrument inertia. When the complex fluid is very soft and the rheological characterization is time dependent, such as in oscillatory tests, instrument inertia artifacts may appear. Microrheology techniques, thanks to the low inertia of colloidal probe particles, allow performing measurement of the linear rheology up to megahertz frequencies [88].

4. Fluid inertia. The assumption of having simple shear flow can be transgressed due to fluid inertia. When characterizing under simple shear soft low-viscous materials, either in transient or steady state conditions, fluid inertia issues may arise. Ewoldt et al. [26] highlighted two major sources of errors in the rheological characterization due to the inertia of the fluid: in oscillatory tests, the combination of high frequencies and low viscoelastic modulus may lead to propagating waves from either viscous momentum diffusion or elastic shear waves or both; secondary flows, either by inertial instabilities or elastic instabilities, may be originated at high velocities. Both fluid inertia effects that mask the physics of interest can be mitigated by performing a rheological characterization at microscale [81] .

To fully describe the flow properties of a complex fluid we must complement shear with extensional rheometry. Despite the recognized importance, the rheological characterization under pure extensional flows has been traditionally less explored due to intrinsic difficulties associated with imposing shear-free conditions. Consequently, the development of instrumentation has been delayed compared to the shear rheometry and mainly focused on high-viscous materials, i.e. polymer melts. Despite the great advances in extensional rheometry of low-viscosity complex fluids during the last years [24, 78, 85], two different filament stretching methods are still dominating the market place (CaBER™and FiSER™). However, as in shear rheometry, some challenges and limitations arise when characterizing the rheological properties of complex fluids in extensional flow:

1. Gravitational forces. Gravitational forces my have undesirable influence on reliability of the measurements obtained in both filament stretching rheometers. Firstly, the interfacial forces due to the surface tension must be larger than the gravitational body forces in order to allow the fluid forming the required liquid bridge between the plates and perform the experiment adequately. Secondly, gravitational forces may pull down the fluid and break the symmetry of the filament during the experiment run, leading to misleading dataset [35]. In microfluidic flows gravitational body forces can be neglected, as the Bond number, establishing the relationship between gravitational body forces and surface tension, is intrinsically very small ($Bo = \frac{\rho L^2 g}{\sigma} \ll 1$) [69].

2. Inertial effects. Despite the CaBER device allows characterizing the extensional properties of viscoelastic liquids with lower viscosity and elasticity than the FiSER, and the combination of the Slow Retraction Method [12] with the use of high speed cameras and small plates [80] may improve the results, inertial effects can be unavoidable below critical limits in viscosity and elasticity [74]. As discussed above, inertial effects can be neglected in rheometry at microscale.

3. Evaporation. In both extensional rheometers, CaBER™and FiSER™, fluid samples enjoy of having free surfaces, which is positive in the sense of having zero parallel shear stresses (shear-free flow), but can be negative when dealing with fluids exhibiting rapid solvent evaporation [71]. In this latter case, the use of microfluidics avoid that practical disadvantage.

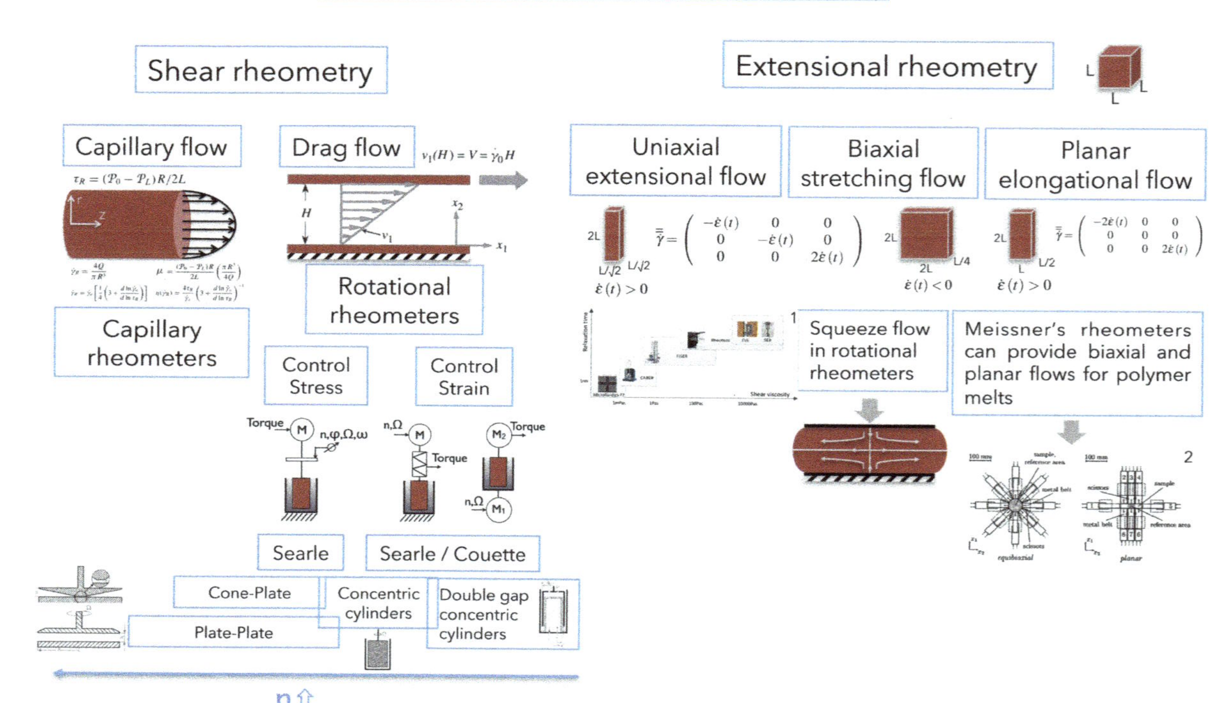

Fig. 1.3 Summary chart of the most common techniques to perform a rheological characterization at macroscale with commercial rheometers. The inset Fig. 1.1 showing the commercially available uniaxial extensional rheometers has been reproduced from Ref. [32] with permission of Springer. The inset Fig. 1.2 has been reprinted with permission from [44], Copyright 2003, The Society of Rheology

"Performing rheometry at microscale can solve many of limitations associated with standard rheometry at macroscale, particularly when working with complex fluids that are soft, have low-viscosity or are prone to evaporation."

1.3 Rheometry at Microscale

In the words of Clasen and McKinley [18] the term *microrheometry* "may be best defined as the science of measuring, quantitatively, the rheological properties of a fluid sample when one characteristic dimension is on the scale of order microns". They also proposed that the variety of different approaches to microrheometry can be sorted out into four major groups, graphically depicted in Fig. 1.4:

1. **Microrheology** allows the determination of the rheological properties of a complex fluid from the motion of probe particles embedded within it. Regarding the source of the force moving the probe particles, it can be distinguished passive microrheology, where particles move due to thermal energy [15, 53], from active microrheology, in which the probes are moved by optical or magnetic tweezers [73, 84]. These techniques are preferred for obtaining local and bulk shear viscoelastic properties both inside and outside the linear viscoelastic region. The reader interested in this topic can find more information in the following references [17, 21, 52, 82].

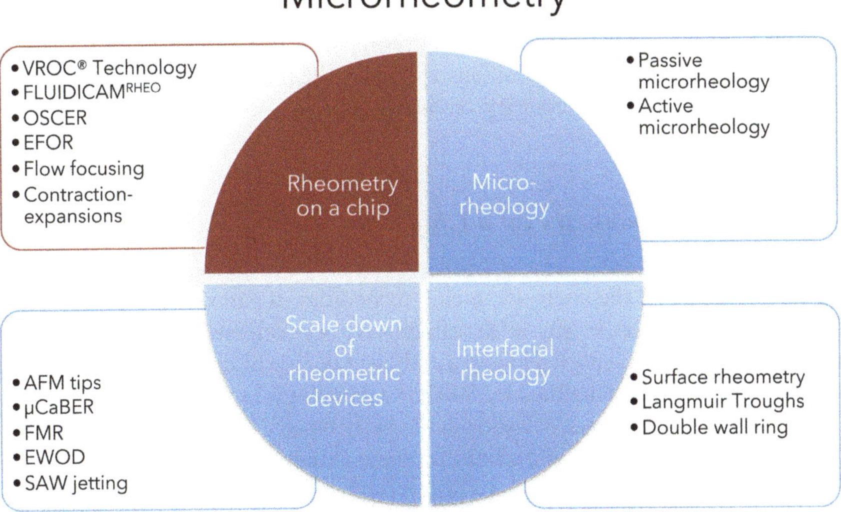

Fig. 1.4 Different approaches to microrheometry sorted out into four major groups

2. **Interfacial rheology** can be considered a kind of microrheometry, as it deals with the characterization of the static and dynamic properties of complex fluid-fluid interfaces, with one characteristic dimension on the scale of order of microns [18]. Complex fluid-fluid interfaces show nonlinear mechanical response to flow and deformation due to the presence of a microstructure, which can be originated by the presence of either surface-active molecules or micro/nano-particles at the interface. Further reading in this field of rheology is highly recommended through the review paper of Fuller and Vermant [28, 29] and their research works, such as [72, 76, 86, 87].

3. **Scale down of rheometric devices** can be of help for minimizing to gravitational and inertial effects inherent to low viscosity complex fluids. It can be found in the literature some examples able to extend the limits of reliable measurements compared with their macroscopic counterparts [9, 18, 25, 57, 80]. However, scaling down of some mechanical subsystems, mainly those including torsional motors, moving parts, and torque transducers may be impractical [69].

4. **Rheometry-on-a-chip** consists of characterizing the rheological properties of complex fluids in a microfluidic channel, under both standard rheometric flows (shear and extensional), and it has already proven great potential, particularly for low-viscosity complex fluids [32, 49, 69]. There are some intrinsic characteristics of microfluidics that turn it a great platform for rheometry: (a) Enhanced elastic behaviour of the fluid due to small length scale; (b) no evaporation issues; (c) small volume sample; (d) negligible inertial or gravitational effects; (e) the potentiality of producing highly integrated, portable and disposable/recyclable devices; (f) the possibility of optimizing numerically the geometry; (g) optical access and a direct characterization of the fluid-flow dynamics (see Chap. 3 for more details), measurement of velocity fields (particle tracking or micro-particle image velocimetry techniques) and stress tensors fields (birefringence technique); and (h) suitable to be used as an online rheological sensor in industrial processes. Hereafter attention will be devoted to the rheometry-on-a-chip approach, as highlighted in Fig. 1.4.

1.3.1 Shear Rheometry on a Chip

The development of rheometry-on-a-chip technology is following the steps of its macroscale counterpart. Firstly, all the efforts have been oriented towards performing steady state measurements, leading to a sufficient degree of matureness to reach the marketplace, as it is exemplified by VROC® technology and Fluidicam^{Rheo} device. The measurement of the viscoelastic moduli in rheometry-on-a-chip approach is yet scarce [16] and more work is required for the commercialization of a microdevice able to provide viscoelastic moduli of complex fluids. At the current state of the art, these sort of characterization is preferred to be performed by microrheology (either active or passive).

> Current commercially available devices performing shear rheometry on a chip can only provide the apparent viscosity curve, that is the dependence of the viscosity with the imposed shear rate. If you are interested in characterizing the viscoelasticity in microscale, you are recommended to perform either passive or active microrheological experiments.

1.3.1.1 VROC® technology

The principle of VROC® technology [4, 68, 69] for performing shear rheometry on a chip (Fig. 1.5a) is based on the fundamentals of capillary rheometry at macroscale for a planar slit geometry, where the viscosity of the sample (η) is obtained from the steady fully developed flow condition by a relationship between the imposed flow rate (Q), the measured pressure drop (ΔP) along a capillary length (L) and a numerical factor (k) depending on the aspect ratio (width/depth, if $w/d \gg 1$ then $k = 6$, while if $w/d = 1$ then $k = 14.3$) of the slit channel [69]. Equation 1.3 indicate that relationship for a laminar Newtonian flow in a planar slit geometry:

$$\tau_{wall} = \frac{wd}{2L\,(w+d)}\Delta P \tag{1.1}$$

$$\dot{\gamma}_{wall} = k\frac{Q}{wd^2} \tag{1.2}$$

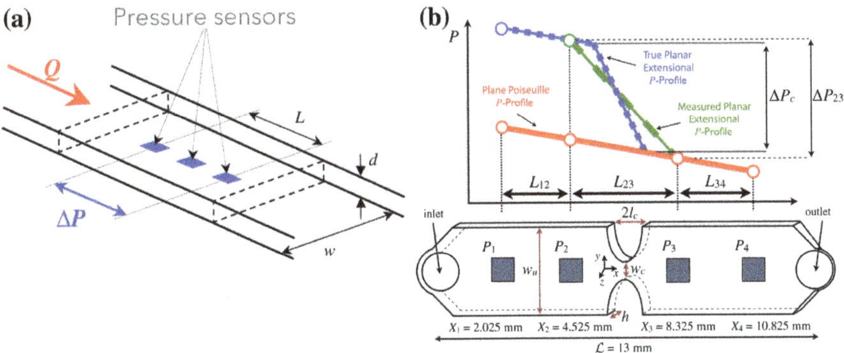

Fig. 1.5 Principle of the VROC® technology developed and commercialized by RheoSense. Reproduced from Ref. [59, 68] with permission of Springer

$$\eta = \frac{\tau_{wall}}{\dot{\gamma}_{wall}} = \frac{w^2 d^3}{2L\,(w+d)\,k}\frac{\Delta P}{Q}, \tag{1.3}$$

For fluids with a shear rate dependent viscosity, the Weissenberg-Rabinowitsch-Mooney correction for rectilinear channels is used to calculate the viscosity [51] and the viscosity is now given by the Eq. 1.7.

$$\tau_{wall} = \frac{wd}{2L\,(w+d)}\Delta P \tag{1.4}$$

$$\dot{\gamma}_{wall-app} = k\frac{Q}{wd^2} \tag{1.5}$$

$$\dot{\gamma}_{wall} = \dot{\gamma}_{wall-app}\frac{1}{3}\left[2 + \frac{dln\dot{\gamma}_{wall-app}}{dln\tau_{wall}}\right] \tag{1.6}$$

$$\eta = \frac{\tau_{wall}}{\dot{\gamma}_{wall}} = \frac{3w^2 d^3}{2L\,(w+d)\left[2 + \frac{dln\dot{\gamma}_{wall-app}}{dln\tau_{wall}}\right]k}\frac{\Delta P}{Q} \tag{1.7}$$

Flush mounted pressure transducers on the wide side of the channel provide direct measurement of the pressure gradient with negligible disturbance of the flow. The pressure transducer must be carefully calibrated to ensure accuracy in the measurements of viscosity over a wide shear rate range, at least an order of magnitude greater than in the standard macroscale rheometers [68]. RheoSense[1] is currently commercializing the VROC® technology, adapted for different samples, working temperatures, etc. The rectangular cross-section microchannel is made of glass and a Si low profile pressure sensor array is flush-mounted on one of the two wide sides. The equipments are provided with a software applications having implemented Eq. 1.7, in which the term $\frac{dln\dot{\gamma}_{wall-app}}{dln\tau_{wall}} = 1$ for Newtonian fluids and Eq. 1.7 then is equal to Eq. 1.3, while for non-Newtonian fluids $\frac{dln\dot{\gamma}_{wall-app}}{dln\tau_{wall}} \neq 1$.

Measuring accurately pressure drop in microfluidic devices is challenging (see Chap. 3 for further information). It can be found in the literature other approaches for measuring the pressure drop, such as the one proposed in [66] for PDMS microchannels. Nevertheless, the flush-mounted approach is probably a robust and adequate approach for a commercial rheometer-on-a-chip device.

> VROC® Technology allows obtaining the viscosity curve of complex liquids (min volume 12 µl) with a shear viscosity ranging from 0.2 to 80,000 mPa·s under shear rates ranging from 0.5 to 1,000 s^{text-1}. Temperature control between 4 and 70 °C.

---

[1] http://www.rheosense.com Cited 30 April 2017.

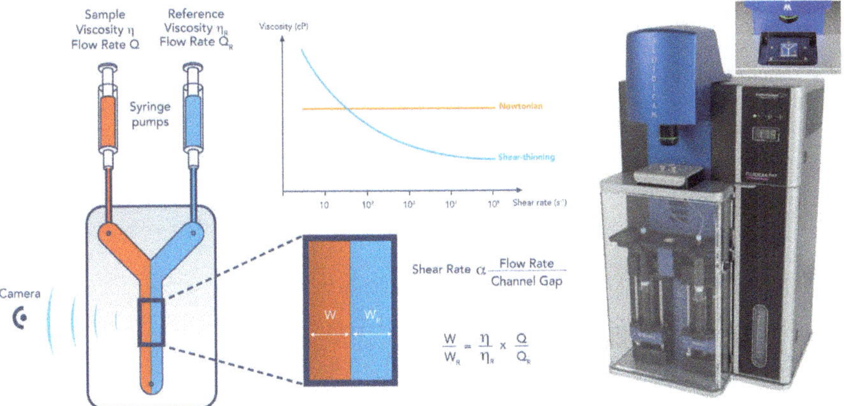

Fig. 1.6 Fluidicam^{Rheo} device by formulaction. Reproduced from Ref. [27] with permission of Formulaction

1.3.1.2 Fluidicam^{Rheo}

The French company Formulaction[2] introduces Fluidicam^{Rheo} in 2016 (Fig. 1.6), a microfluidic rheometer based on live imaging of a microfluidic co-flow. The technology was first demonstrated by Colin and co-workers [19, 42, 43], inspired by the work of Galambos and Foster [30].

The principle is straightforward: two fluids, the sample and a Newtonian reference fluid, flows together in a microfluidic channel. As the flow is laminar, a visible interface is formed between both fluids (miscible or not). Shear rate and sample viscosity is simply calculated from the knowledge of the reference viscosity, geometries of the fluid streams and flow rates. More precisely:

- Pressure gradient ∇P_r is calculated in the reference stream, knowing width W_r (measured with the camera), flow rate Q_r, viscosity η_r and channel depth h, using well established relationships [11], where suffixes r refer to the *reference fluid*. Considering $h \ll W_r$, pressure gradient can be written as $\nabla P_r = -\frac{12\eta_r}{h^3 W_r} Q_r$. More accurate models can also be used, better describing the flow in a rectangular cross section [75].
- Interface being straight, parallel to channel side walls, it is assumed that the pressure gradient in the sample stream equals the pressure gradient in the reference stream (otherwise there would be a visible transverse flow): $\nabla P = \nabla P_r$.
- Shear stress at the walls σ_w (up or bottom, separated by the channel depth h in Fig. 1.7) is calculated from the pressure gradient in the sample stream (balance between surface shear stresses and body pressure forces acting over a fluid element) [51]: $\sigma_w = -\frac{h}{2}\nabla P$.

---

[2]http://www.formulaction.com Cited 30 April 2017.

Fig. 1.7 Fluidicam^{Rheo} co-flow principle. Reproduced from Ref. [27] with permission of Formulaction

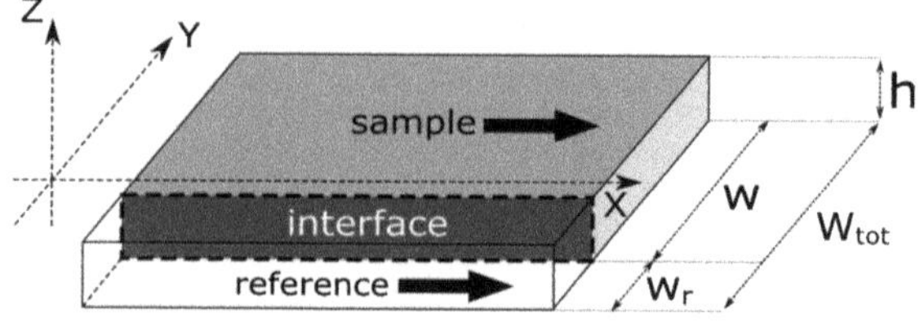

- Shear rate at the wall is calculated using the Rabinowitsch equation, independently from the rheological model: $\dot{\gamma}_w = \frac{2Q}{wh^2}\left[2 + \frac{dlnQ}{dln\sigma_w}\right]$.
- Finally, viscosity at the wall is calculated from its definition: $\eta_w = \frac{\sigma_w}{\dot{\gamma}_w}$.

In the original work from Colin et al. [19, 42, 43], the method involves manual settings of the flow rates until the interface is visible and located in a position suitable for measurement with a good accuracy. This step requires not only user attendance, but is also both time and sample consuming. Moreover, shear rates was not known before analysis, as it value depends on the interface position. This is an issue as most rheologists prefer to drive their viscosity measurements over a range of shear rates (representative of the end-user application) defined before the experiment, and not over a range of flow rates. Fluidicam^{Rheo} overcomes these issues with a user-friendly software built around a predictive control of the interface position, that allows to control experiments in term of shear rates (instead of flow rates), and a fully automated operation without user attendance. The instrument also integrates a fast temperature control from 4 to 80 °C. Finally, live imaging make understanding of rheological experiments much easier and comfortable, as it become possible to see the product flowing as the measurement is running.

Fluidicam^{Rheo} allows obtaining the viscosity curve of complex liquids (min volume 500 µl) with a shear viscosity ranging from 0.1 to 200,000 mPa·s under shear rates ranging from 100 to 100,000 s^{-1}. Temperature control between 4 and 80 °C.

1.3.1.3 Other Non-commercialized Approaches

Besides the approaches described in Sects. 1.3.1.1 and 1.3.1.2, there are many other approaches for performing shear rheometry on a chip that have not been commercialized yet. The following ones are innovative enough to deserve a mention here:

- **iCapillary-based viscometer**. Solomon et al. [79] recently proposed a new version of an image-based slit-microrheometer (Fig. 1.8). A constant pressure (P_c) is imposed at the inlet of a straight (L_{ch}) rectangular cross-section microchannel

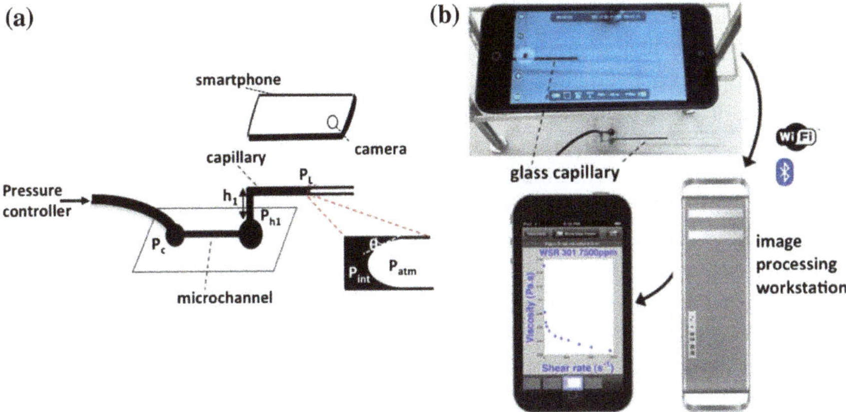

Fig. 1.8 Schematic and operation of the iCapillary-based viscometer. Reproduced from Ref. [79] with permission of Springer

($w \gg h$). A L-shaped millimeter-scale glass capillary is connected at the outlet of the microchannel, having a vertical section h_1 providing a static head P_{h_1}. Thanks to the large diameter glass capillary, a smartphone camera can be used to record the motion of the interfacial meniscus and the capillary pressure due to the air-fluid interface is small compared to the driving pressure ($P_L = \frac{2\sigma cos\theta}{r}$). Thus, the wall shear stress is given by Eq. 1.8.

$$\tau_{wall} = \frac{\left(P_c - P_{h_1} + P_L\right) wh}{2L_{ch}\left(w + h\right)} \tag{1.8}$$

They also consider also Weissenberg-Rabinowitch-Mooney equation for determining the wall shear rate as a function of the flow rate. As the experiments are pressure driven, the flow rate is estimated by measuring the mean velocity of the meniscus (distance traveled over a period of time) and multiplying it by the cross-sectional area of the capillary. Thus, the viscosity is calculated for each imposed inlet pressure as $\eta = \frac{\tau_{wall}}{\dot{\gamma}_{wall}}$.

The main novelty of this approach lies on how τ_{wall} is calculated without using pressure sensors but controlling the pressure at the inlet and the outlet, and how the flow rate is calculated by means of an image analysis approach.

- **T-Junction microdevice.** While the measurement of a viscosity curve traditionally requires of multiple measurements, that is for every different point of the curve a different flow condition must be imposed, the approach proposed by Zimmerman and Ress [90] aims at improving the efficiency by obtaining the whole viscosity curve of a complex fluid in a single experiment. Inducing the flow field in such a microfludic channel either by an applied pressure drop or by an electric field (just for electrokinetic flows), a range of shear rates can be generated within a single experiment due to the gradients on the characteristic velocity field associated with the flow in a T-junction. Bandalusena et al. [5–7] developed a procedure for assessing the total rheometric response for such a system based on the statistical moments of the velocity field (measured by means of μ-PIV technique) analysed in conjunction with finite element modelling. Despite the reliability of μ-PIV technique for providing the velocity field at microscale (Chap. 3), the equipment is expensive and bulky. Zimmerman et al. [89] showed numerically that it also exists a one-to-one mapping between the first three statistical moments of the end wall pressure profile with the parameters of the GNF models. The rheological curves can be also obtained from the pressure field in such a flow, which can be measured by means of piezoelectric pressure transducers in a more cost-effective way. Thus, from measurements from flow and pressure sensors in a microchannel T-junction, it is possible to solve the mathematically proscribed'inverse problem' in order to obtain the constitutive viscous parameters for GNF fluids. The addition of transient effects caused by pulsing of the fluid would allow to obtain the constitutive viscous parameters for viscoelastic fluids [90]. This latter feature is a step forward and major distinction regarding the previously described shear-rheometry-on-a-chip approaches.

> Zimmerman and Ress approach would not only allow to obtain the constitutive viscous parameter for GNF fluids running just one experiment at steady flow conditions, but also the constitutive parameters for viscoelastic fluids under transient flow conditions.

1.3.2 Extensional Rheometry on a Chip

As in macroscale rheometry, despite the importance of characterizing the extensional properties of complex fluids, the development of experimental techniques at microscale for performing extensional flow tests has been delayed regarding the development of the shear techniques. To the best knowledge of the author, there is only one commercially available extensional rheometer-on-a-chip, based on VROC® technology.

1.3.2.1 EVROC™ viscometer

eVROC™ viscometer is based on Rheosense's patented VROC® technology [4], previously described in Sect. 1.3.1.1 for shear rheometry on a chip. In that case, the geometry microfluidic channel was straight and with a rectangular cross section, like the macroscale counterpart for performing capillary experiments. That technology can be adapted for measuring both the shear and extensional viscosities simultaneously.

The chip used in eVROC^{TM} has a hyperbolic contraction/expansion zone in the middle of the microchannel and four monolithically integrated MEMS pressure sensors symmetrically distributed upstream and downstream the contraction throat[3] (Fig. 1.5b). While abrupt contractions fail to produce homogeneous extension conditions [64], hyperbolically-shaped contraction in planar configurations allow to impose a strong extensional flow with an approximately constant extension rate along its centerline of the microchannel [13, 49, 59, 63], although there is an important contribution of the shear flow close to the lateral walls. This latter issue is the main reason why the eVROC^{TM} microdevice includes four pressure sensors, for evaluating separately the pressure drop due to the viscous shear stresses (ΔP_v) in the fully developed regions (upstream and downstream the throat) and the pressure drop across the contraction/expansion zone (ΔP_c). By subtracting them, it would be possible to evaluate the pressure drop associated with the pure extensional flow ($\Delta P_e = \Delta P_c - \Delta P_v$) and resulting from the elastic normal stresses alone [59]. Thus, the apparent extensional viscosity for fully developed extensional flow in the hyperbolic contraction is given by Eq. 1.9:

$$\eta_{E,a} = \frac{1}{\varepsilon_H} \frac{\Delta P_e}{\dot{\varepsilon}_a}, \tag{1.9}$$

where $\dot{\varepsilon}_a = \frac{Q}{l_c h}\left(\frac{1}{w_c} - \frac{1}{w_u}\right)$ is the apparent extension rate, as a function of the flow rate (Q) and the geometrical parameters defining the hyperbolic contraction/expansion, and $\varepsilon_H = \int_0^t \dot{\varepsilon}_a dt'$ is the Hencky strain experience by a fluid element [59].

> eVROC^{TM} viscometer allows measuring the apparent extensional viscosity of complex liquids with a shear viscosity ranging from 20 to 2000 mPa·s under extension rates ranging from 0.1 to 1000 s^{-1}.

1.3.2.2 Other Non-commercialized Approaches

As in Sect. 1.3.1.3, there are also many other approaches in the literature for performing extensional rheometry on a chip that have not been commercialized yet. These alternative approaches look for technique able to overcome the major limitation of eVROC^{TM} viscometer, which is the fact of having a region with combined shear and elongated characteristics in the flow kinematics [91]. The following ones are innovative enough to deserve a mention here:

---

[3]http://www.rheosense.com.

- **Optimized geometries**. In order to overcome the limitation of having a non-homogeneous velocity profile which results in a constant extension rate, Alves [1] developed an algorithm for optimal design of the geometry of the microchannel so that it provides the desired velocity field. That algorithm has been successfully implemented for different microfluidic devices, i.e. 2D [48] and 3D [34] cross-slots, hyperbolic contraction/expansion [34] and T-junction [2] providing uniform extension rates profiles. Figure 1.9 shows the optimization flowchart, a numerically optimized design, and the target velocity and strain-rate profiles along the centerline for a 3D optimized cross-slots geometry.

> The combination of numerical flow simulations (see Chap. 4), numerical optimization techniques (see Chap. 5) and the experimental characterization fluid-flow dynamics (see Chap. 3) has been proven to provide excellent results for undergoing extensional rheometry on a chip.

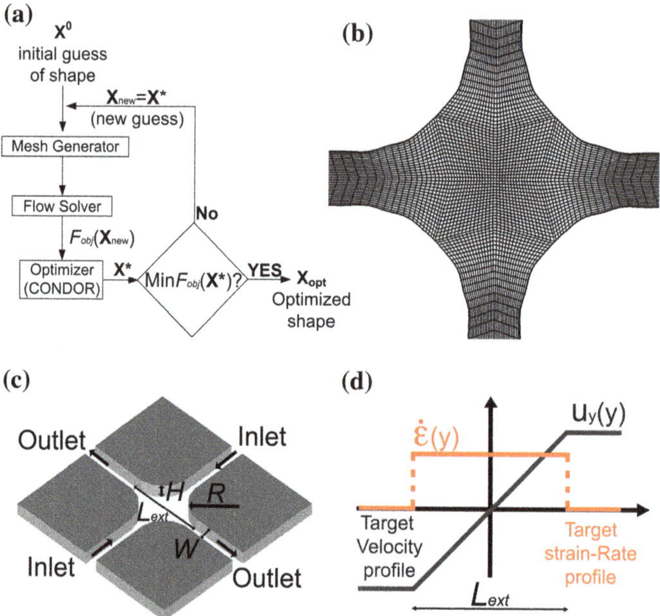

Fig. 1.9 Optimal shape design procedure. **a** Schematic illustration of the optimization flowchart, **b** top-view of an exemplifying optimized design and corresponding mesh, **c** 3D illustration of the optimized design including the geometric parameters, **d** target velocity and strain-rate profiles along the vertical centreline (x = 0). Reproduced from Ref. [34] with permission from The Royal Society of Chemistry

Working with these optimized geometries and the currently available experimental techniques for the characterization of the fluid-flow dynamics at microscale (see Chap. 3), it is possible to determine the extensional viscosity of low viscosity complex fluids accurately.

- **Particle migration**. Del Giudice et al. [23] proposed a completely different approach to determine the relaxation time of viscoelastic fluids (on the order of milliseconds) in a straight microfluidic channel, based on the viscoelasticity-induced particle migration phenomenon. The migration of a suspended particle in a flowing viscoelastic fluid towards the center of the microchannel is not due to inertia-driven migration effects, but instead it is governed by the dimensionless parameter $\theta = De\beta^2 \frac{L}{H}$, being L the distance from the inlet, H the depth of the channel and $\beta = D_p/H$ the confinement ratio of the particle with a diameter D_p. Recommended ranges for these two parameters are $H = 50 - 100$ μm and $\beta = 0.05 - 0.1$. Thus, by means of particle tracking experiments, it is calculated the particle distribution in the first band f_1 is calculated at different flow rates. Subsequently, θ is computed from f_1, and from that it is obtained the value of the Deborah number, providing the measurement of the fluid relaxation time. As θ is inversely proportional to H^4, very low relaxation times can be measured by using smaller channels, with a remarkable increase in the sensitivity of the measurements.

When performing rheometry on a chip, either under shear or extensional flow, it is important to adequately select the materials and the experimental tools. For instance, high pressures may deform the cross section of the PDMS microchannels produced by soft lithography techniques [45], endangering the reliability of the measurements and limiting the maximum values of shear rates or extension rates. In these cases, it is recommended the use of rigid materials (stainless steel, PMMA, fused silica, borosilicate glass, etc.) instead (see Chap. 2 for further information about the different fabrication techniques). Additionally, if the measurement of the pressure drop is relevant for the rheological characterization, it must be assessed if we should go for flush mounted or pressure taps (see Chap. 3 for further information about the different fluid-flow characterization techniques).

1.3.3 On-line Rheology Sensor: RheoStream®

One key aspect on industrial manufacturing of complex fluids (paints, food and beverages, pharmaceutical products, detergents, etc.) is monitoring the rheological properties of the fluid at different stages of the production in real time, so that the process operators and control systems can make decisions about their production. Traditionally, the viscosity was the fluid property measured on real time by means of different range of commercially available in-line viscometers. Nevertheless, as we already know, viscosity is not enough information when dealing with the process of complex fluids. Standard rheometry at macroscale can provide a full rheological

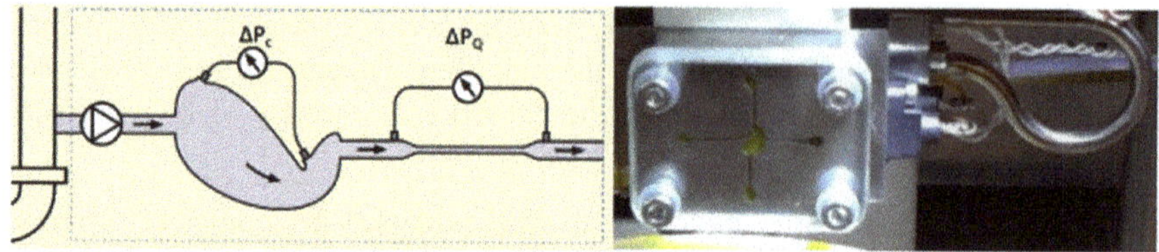

Fig. 1.10 RheoStream®, working principle (*left*) and a picture of the device (*right*). Reproduced with permission of Fluidan

characterization of these fluids in the laboratory, that is out-line of the production process. This implies a delay that disable the real-time control of the production process and that may result in a bottle neck in development, production and quality control. Thus, monitoring key rheological properties of complex fluids in-line the production process is crucial for manufacturing non-Newtonian fluids.

The RheoStream® device is commercialized by Fluidan[4] and can provide real-time continuous key rheological parameters in a given product or process. This technology combines capillary viscometry with a topologically optimized flow cell in which the rheological properties are related with differential pressure generated by the fluid-flow through it (Fig. 1.10) [61, 62]. The fundamental idea is that the total pressure drop measured in the optimized cell (ΔP_c) is the sum of two contributions: (1) the viscous contribution ($\Delta P_{c,V}$), which can be inferred from the capillary measurement ($\Delta P_{c,V} \propto \Delta P_Q$); and (2) the viscoelastic one ($\Delta P_{VE} = \Delta P_c - \Delta P_{c,V}$), which is calculated. The first normal stress in the fluid is proportional to the viscoelastic contribution to the pressure drop in the cell ($N_1 = k_1 \Delta P_{VE}$) [3] and N_1 can be correlated to the elastic modulus (G') by means of the extended Cox-Merz rule ($G' = C N_1^{\alpha} \dot{\gamma}_c^{2(1-\alpha)}$) [83], where C and α are calibration constant that must be determined for each working fluid, and $\dot{\gamma}_c$ is the characteristic shear rate in the flow cell that is correlated to the frequency in the oscillatory measurement performed in a rotational rheometer. Once the elastic modulus is obtained, the viscous modulus can be calculated ($G'' = \sqrt{G^{*2} - G'^2}$), where G^* is the complex modulus estimated from rescaling the shear stress (σ) by applying Cox-Merz rule [20].

The true power of RheoStream® is providing in-line real-time rheological parameters of complex fluids. Nevertheless, these parameters are not measured from "first principle", as the instrument needs to be calibrated to the working fluid.

---

[4]http://www.fluidan.com Cited 30 April 2017.

1.4 Summary

In this chapter, the following 10 key remarks/tips have been provided:

#1 GNF's are always be considered as inelastic fluids. Only VEF's exhibit simultaneously properties of elastic solids and viscous liquids, and therefore a characteristic relaxation time (λ).

#2 Even for fluids with small relaxation times, microfluidics allows the exploration of zones in the Wi-Re parameter space unreachable at macroscale, due to the very small characteristic length scale that enhances the elastic effects ($L \downarrow \Rightarrow El \uparrow$).

#3 Performing rheometry at microscale can solve many of limitations associated with standard rheometry at macroscale, particularly when working with complex fluids that are soft, have low-viscosity or are prone to evaporation.

#4 Current commercially available devices performing shear rheometry on a chip can only provide the apparent viscosity curve, that is the dependence of the viscosity with the imposed shear rate. If you are interested in characterizing the viscoelasticity in microscale, you are recommended to perform either passive or active microrheological experiments.

#5 VROC® Technology allows obtaining the viscosity curve of complex liquids (min volume 12 μl) with a shear viscosity ranging from 0.2 to 80,000 mPa·s under shear rates ranging from 0.5 to 1,000 s^{-1}. Temperature control between 4 and 70 °C.

#6 Fluidicam^{Rheo} allows obtaining the viscosity curve of complex liquids (min volume 500 μl) with a shear viscosity ranging from 0.1 to 200,000 mPa·s under shear rates ranging from 100 to 100,000 s^{-1}. Temperature control between 4 and 80 °C.

#7 Zimmerman and Ress approach would not only allow to obtain the constitutive viscous parameter for GNF fluids running just one experiment at steady flow conditions, but also the constitutive parameters for viscoelastic fluids under transient flow conditions.

#8 eVROC^{TM} viscometer allows measuring the apparent extensional viscosity of complex liquids with a shear viscosity ranging from 20 to 2000 mPa·s under extension rates ranging from 0.1 to 1000 s^{-1}.

#9 The combination of numerical flow simulations, numerical optimization techniques and the experimental characterization fluid-flow dynamics has been proven to provide excellent results for undergoing extensional rheometry on a chip.

#10 The true power of RheoStream® is providing in-line real-time rheological parameters of complex fluids. Nevertheless, these parameters are not measured from "first principle", as the instrument needs to be calibrated to the working fluid.

Acknowledgements F.J. Galindo-Rosales would like to acknowledge the financial support from FCT, COMPETE and FEDER through grant IF/00190/2013 and project IF/00190/2013/CP1160/CT 0003. The author would also like to thank Formulaction and Fluidan for fruitful discussion about their products, and for graciously providing the pictures used in Figs. 1.6, 1.7, and 1.10, respectively.

References

1. Alves, M. A. (2008). Design of a Cross-Slot flow channel for extensional viscosity measurements. *AIP Conference Proceedings, 1027*(1), 240–242.
2. Alves, M. A. (2011). Design of optimized microfluidic devices for viscoelastic fluid flow. Technical Proceedings of the 2011 NSTI Nanotechnology Conference and Expo. 2: 474–477.
3. Baird, D. G. (2008). First normal stress difference measurements for polymer melts at high shear rates in a slit-die using hole and exit pressure data. *Journal of Non-Newtonian Fluid Mechanics, 148*, 13–23.
4. Baek, S. G. (2010). Micro rheometer for measuring flow viscosity and elasticity for micron sample volumes. US Patent 7,770,436.
5. Bandalusena, H. C. H., Zimmerman, W. C., & Rees, J. M. (2009). Microfluidic rheometry of a polymer solution by micron resolution particle image velocimetry: A model validation study. *Measurement Science and Technology, 20*(11), 115404.
6. Bandulasena, H. C. H., Zimmerman, W. B., & Rees, J. M. (2010). Creeping flow analysis of an integrated microfluidic device for rheometry. *Journal of Non-Newtonian Fluid Mechanics, 165*, 1302–1308.
7. Bandulasena, H. C. H., Zimmerman, W. B., & Rees, J. M. (2010). Rheometry of non-Newtonian polymer solution by microchannel pressure driven flow. *Applied Rheology, 20*, 55608.
8. Barnes H. A., Hutton J. F., & Walters, K. (1993). An introduction to Rheology. Rheology Series (Vol. 3). Amsterdam: Elsevier Science Publishers B.V.
9. Bhattacharjee, P. K., McDonnell, A. G., Prabhakar, R., Yeo, L. Y., & Friend, J. (2011). Extensional flow of low-viscosity fluids in capillary bridges formed by pulsed surface acoustic wave jetting. *New Journal of Physics, 13*, 023005.
10. Bird, R. B., Armstrong, R. C., & Hassager, O. (1987). Dynamics of polymer liquids. Volume 1 - Fluid mechanics (2 edn). New York: Wiley.
11. Bruus, H. (2007). Theoretical microfluidics. Oxford University Press.
12. Campo-Deaño, L., & Clasen, C. (2010). The slow retraction method (SRM) for the determination of ultra-short relaxation times in capillary breakup extensional rheometry experiments. *Journal of Non-Newtonian Fluid Mechanics, 165*(2324), 16881699.
13. Campo-Deaño, L., Galindo-Rosales, F. J., Oliveira, M. S. N., Alves, M. A., & Pinho, F. T. (2011). Flow of low viscosity boger fluids through a microfluidic hyperbolic contraction. *Journal of non-Newtonian Fluid Mechanics, 166*(21–22), 1286–1296.
14. Campo-Deaño, L., Galindo-Rosales, F. J., Pinho, F. T., Alves, M. A., & Oliveira, M. S. N. (2012). Nanogel formation of polymer solutions flowing through porous media. *Soft Matter, 8*(24), 6445–6453.
15. Campo-Deaño, L., Dullens, R. P. A., Aarts, D. G. A. L., Pinho, F. T., & Oliveira, M. S. N. (2013). Viscoelasticity of blood and viscoelastic blood analogues for use in polydymethylsiloxane in vitro models of the circulatory system. *Biomicrofluidics, 7*, 034102.
16. Christopher, G. F., Yoo, J. M., Dagalakis, N., Hudson, S. D., & Migler, K. B. (2010). Development of a MEMS based dynamic rheometer. *Lab Chip, 10*, 2749–2757.
17. Cicuta, P., & Donald, A. M. (2007). Microrheology: a review of the method and applications. *Soft Matter, 3*, 1449–1455.
18. Clasen, C., & McKinley, G. H. (2004). Gap-dependent microrheometry of complex liquids. *Journal of non-Newtonian Fluid Mechanics, 124*(1), 1–10.
19. Colin, A., Cristobal, G., Guillot, P., & Joanicot, M. (2012). Method and installation for determining rheological characteristics of a fluid, and corresponding identifying method. US Patent 8,104,329.
20. Cox, W. P., & Merz, E. H. (1958). Correlation of dynamic and steady flow viscosities. *Journal of Polymer Science, 28*, 619–622.
21. Crocker, J. C., Valentine, M. T., Weeks, E. R., Gisler, T., Kaplan, P. D., Yodh, A. G., et al. (2000). Two-point microrheology of inhomogeneous soft materials. *Physical Review Letters, 85*, 888.

22. Dealy, J. M. (2010). Weissenberg and deborah numbers their definition and use. *Rheology Bulletin (The Society of Rheology)*, *79*(2), 14–18.
23. Del Giudice, F., D'Avino, G., Greco, F., De Santo, I., Nettiab, P. A., & Maffettone, P. L. (2015). Rheometry-on-a-chip: measuring the relaxation time of a viscoelastic liquid through particle migration in microchannel flows. *Lab on a Chip*, *15*, 783–792.
24. Dinic, J., Zhang, Y., Jimenez, L. N., & Sharma, V. (2015). Extensional relaxation times of dilute, aqueous polymer solutions. *ACS Macro Letter*, *4*, 804808.
25. Erni, P., Varagnat, M., Clasen, C., Crest, J., & McKinley, G. H. (2011). Microrheometry of sub-nanolitre biopolymer samples: non-newtonian flow phenomena of carnivorous plant mucilage. *Soft Matter*, *7*, 10889.
26. Ewoldt, R. H., Johnston, M. T., & Caretta, L.M. (2015). Complex fluids in biological systems: experiment, theory, and computation. In: Spagnolie, S.E., (ed.) Chapter 6: experimental challenges of shear rheology: how to avoid bad data. New York: Springer.
27. Formulaction (2017). Fluidicam^{Rheo} Technical specifications.
28. Fuller, G. G., & Verman, J. (2011). Editorial: dynamics and rheology of complex fluid-fluid interfaces. *Soft Matter*, *7*(17), 7583–7585.
29. Fuller, G. G., & Verman, J. (2012). Complex fluid-fluid interfaces: rheology and structure. *Annual Review of Chemical and Biomolecular Engineering*, *3*(1), 519–543.
30. Galambos, P., & Foster, F. (1998). An optical micro-fluidic viscometer. DSC-Vol. 66, Micro-Electro-Mechanicical System (MEMS). ASME International Mechanical Engineering Congress and Exposition, 15–20, Anaheim, CA.
31. Galindo-Rosales, F. J., Campo-Deaño, L., Pinho, F. T., van Bokhorst, E., Hamersma, P. J., Oliveira, M. S. N., et al. (2012). Microfluidic systems for the analysis of viscoelastic fluid flow phenomena in porous media. *Microfluidics and Nanofluidics*, *12*(1), 485–498.
32. Galindo-Rosales, F. J., Alves, M. A., & Oliveira, M. S. N. (2013). Microdevices for extensional rheometry of slow viscosity elastic liquids: a review. *Microfluidics Nanofluidics*, *14*, 1–19.
33. Galindo-Rosales, F. J., Campo-Deaño, L., Sousa, P. C., Ribeiro, V. M., Oliveira, M. S. N., Alves, M. A., et al. (2014). Viscoelastic instabilities in micro-scale flows. *Experimental Thermal and Fluid Science*, *59*, 128139.
34. Galindo-Rosales, F. J., Oliveira, M. S. N., & Alves, M. A. (2014). Optimized cross-slot microdevices for homogeneous extension. *RSC Advances*, *4*(15), 7799–7804.
35. Galindo-Rosales, F. J., Segovia-Gutirrez, J. P., Pinho, F. T., Alves, M. A., & de Vicente, J. (2015). Extensional rheometry of magnetic dispersions. *Journal of Rheology*, *59*(1), 193–209.
36. Galindo-Rosales, F. J., Martínez-Aranda, S., & Campo-Deaño, L. (2015). CorkSTFμfluidics - A novel concept for the development of eco-friendly light-weight energy absorbing composites. *Materials & Design*, *82*, 326–334.
37. Galindo-Rosales, F. J., & Campo-Deaño, L. (2016). Composite layer material for dampening external load, obtaining process, and uses thereof. WO Patent App. PCT/IB2015/057,399.
38. Galindo-Rosales, F. J. (2016). Complex fluids in energy dissipating systems. *Applied Sciences*, *6*(8), 206.
39. Groisman, A., & Steinberg, V. (2001). Efficient mixing at low Reynolds numbers using polymer additives. *Nature*, *410*, 905–908.
40. Groisman, A., Enzelberger, M., & Quake, S. R. (2003). Microfluidic memory and control devices. *Science*, *300*(5621), 955–958.
41. Groisman, A., & Quake, S. R. (2004). A microfluidic rectifier: anisotropic flow resistance at low reynolds numbers. *Physical Review Letters*, *92*(2), 094501.
42. Guillot, P., Panizza, P., Salmon, J.-B., Joanicot, M., & Colin, A. (2006). Viscosimeter on a microfluidic chip. *Langmuir*, *22*, 6438–6445.
43. Guillot, P., Moulin, T., Kotitz, R., Guirardel, M., Dodge, A., Joanicot, M., et al. (2008). Towards a continuous microfluidic rheometer. *Microfluidics Nanofluidics*, *5*, 619–630.
44. Hachmann, P., & Meissner, J. (2003). Rheometer for equibiaxial and planar elongations of polymer melts. *Journal of Rheology*, *47*(4), 989–1010.
45. Hardy, B. S., Uechi, K., Zhen, J., & Kavehpour, H. P. (2009). The deformation of flexible PDMS microchannels under a pressure driven flow. *Lab on a Chip*, *9*, 935–938.

46. Haward, S. J., Odell, J. A., Berry, M., & Hall, T. (2011). Extensional rheology of human saliva. *Rheologica Acta, 50*(11), 869–879.

47. Haward, S. J., Sharma, V., & Odell, J. A. (2011). Extensional opto-rheometry with biofluids and ultra-dilute polymer solutions. *Soft Matter, 7*(21), 9908–9921.

48. Haward, S. J., Oliveira, M. S. N., Alves, M. A., & McKinley, G. H. (2012). Optimized cross-slot flow geometry for microfluidic extensional rheometry. *Physical Review Letters, 109*(12), 128301.

49. Haward, S. J. (2016). Microfluidic extensional rheometry using stagnation point flow. *Biomicrofluidics, 10*(4), 043401.

50. Larson, R. G. (1999). *The structure and rheology of complex fluids*. New York, United States: Oxford University Press.

51. Macosko, C. W. (1994). *Rheology: principles, measurements, and applications*. United States: Wiley-VCH Inc.

52. Mason, T. G., & Weitz, D. A. (1995). Optical measurements of frequency-dependent linear viscoelastic moduli of complex fluids. *Physical Review Letters, 74*, 1250.

53. Mason, T. G. (2000). Estimating the viscoelastic moduli of complex fluids using the generalized Stokes-Einstein equation. *Rheologica Acta, 39*, 371–378.

54. Mewis, J., & Wagner, N. J. (2012). *Colloidal suspension rheology*. United Kindom: Cambridge University Press.

55. Mezger, T. G. (2002). The Rheology Handbook: for user of rotational and oscillatory rheometers. Vincentz Verlag, Germany.

56. Morrison, F. A. (2001). *Understanding rheology*. United States: Oxford University Press.

57. Nelson, W. C., Kavehpour, H. P., & Kim, C. J. (2011). A miniature capillary breakup extensional rheometer by electrostatically assisted generation of liquid filaments. *Lab Chip, 11*, 2424–243.

58. Nguyen, N.-T. (2012). Micro-magnetofluidics: interactions between magnetism and fluid flow on the microscale. *Microfluidics and Nanofluidics, 12*(1), 1–16.

59. Ober, T. J., Haward, S. J., Pipe, C. J., Soulages, J., & McKinley, G. H. (2013). Microfluidic extensional rheometry using a hyperbolic contraction geometry. *Rheologica Acta, 52*(6), 529–546.

60. Official symbols and nomenclature of the society of rheology (2013). Journal of Rheology 57(4): 1047–1055.

61. Okkels, F. (2012). A flow measurement device and method. WO Patent App. PCT/DK2012/050,21.

62. Okkels, F., Oestergard, A.L., & Mohammadifar, M.A. (2017). Novel method for on-line rheology measurement in manufacturing of non-Newtonian liquids. Annual Transactions of the Nordic Rheology Society (Vol. 25).

63. Oliveira, M. S. N., Alves, M. A., Pinho, F. T., & McKinley, G. H. (2007). Viscous flow through microfabricated hyperbolic contractions. *Experiments in Fluids, 43*(2–3), 437–451.

64. Oliveira, M. S. N., Rodd, L. E., McKinley, G. H., & Alves, M. A. (2008). Simulations of extensional flow in microrheometric devices. *Microfluidics and Nanofluidics, 5*, 809–826.

65. Oliveira, M. S. N., Alves, M. A., & Pinho, F. T. (2011) Transport and mixing in laminar flows: from microfluidics to oceanic currents. Chapter 6: microfluidic flows of viscoelastic fluids. Wiley-VCH Verlag GmbH & Co. KGaA.

66. Pan, L., & Arratia, P. E. (2013). A high-shear, low Reynolds number microfluidic rheometer. *Microfluid Nanofluid, 14*, 885894.

67. Sheng, P., & Wen, W. (2012). Electrorheological Fluids: Mechanisms, Dynamics, and Microfluidics Applications. *Annual Review Fluid Mechanics, 44*, 14374.

68. Pipe, C., Majmudar, T. S., & McKinley, G. H. (2008). High shear rate viscometry. *Rheologica Acta, 47*(5), 621–642.

69. Pipe, C., & McKinley, G. H. (2009). Microfluidic rheometry. *Mechanics Research Communications, 36*, 110–120.

70. Poole, R. J. (2012). The Deborah and Weissenberg numbers. *British Society of Rheology, Rheology Bulletin, 53*(2), 32–39.

71. Regev, O., Vandebril, S., Zussman, E., & Clasen, C. (2010). The role of interfacial viscoelasticity in the stabilization of an electrospun jet. *Polymer, 51*, 2611–2620.

72. Reynaert, S., Brooks, C., Moldenaers, P., Vermant, J., & Fuller, G. G. (2008). Analysis of the magnetic rod interfacial stress rheometer. *Journal of Rheology, 52*(1), 261–285.

73. Rich, J. P., Lammerding, J., McKinley, G. H., & Doyle, P. S. (2011). Nonlinear microrheology of an aging, yield stress fluid using magnetic tweezers. *Soft Matter, 7*(21), 9933–9943.

74. Rodd, L. E., Scott, T. P., Cooper-White, J. J., & McKinley, G. H. (2005). Capillary break-up rheometry of low-viscosity elastic fluids. *Applied Rheology, 15*, 1227.

75. Steinke, M. E., & Kandlikar, S. G. (2006). Single-phase liquid friction factors in microchannels. *International Journal of Thermal Sciences, 45*, 1073–1083.

76. Samaniuk, J. R., & Vermant, J. (2014). Micro and macrorheology at fluid-fluid interfaces. *Soft Matter, 10*(36), 7023–7033.

77. Schramm, G. (2000). *A practical approach to rheology and rheometry.* Karlsruhe, Germany: Haake GmbH.

78. Sharma, V., Haward, S. J., Serdy, J., Keshavarz, B., Soderlund, A., Threlfall-Holmes, P., et al. (2015). The rheology of aqueous solutions of ethyl hydroxy-ethyl cellulose (EHEC) and its hydrophobically modified analogue (hmEHEC): extensional flow response in capillary break-up, jetting (ROJER) and in a cross-slot extensional rheometer. *Soft Matter, 11*, 32513270.

79. Solomon, D. E., Abdel-Raziq, A., & Vanapalli, S. A. (2016). *Rheologica Acta, 55*(9), 727–738.

80. Sousa, P. C., Vega, E. J., Sousa, R. G., Montanero, J. M., & Alves, M. A. (2017). Measurement of relaxation times in extensional flow of weakly viscoelastic polymer solutions. *Rheologica Acta, 56*(1), 1120.

81. Squires, T. M., & Quake, S. R. (2005). Microfluidics: Fluid physics at the nanoliter scale. *Reviews of Modern Physics, 77*, 977–1026.

82. Squires, T. M., & Mason, T. G. (2010). Fluid Mechanics of Microrheology. *Annual Review of Fluid Mechanics, 42*, 413–38.

83. Steffe, J. F. (1996). *Rheological methods in food engineering process* (2nd ed.). Press, East Lansing, USA: Freeman.

84. Tassieri, T., Del Giudice, F., Robertson, E. J., Jain, N., Fries, B., Wilson, R., et al. (2015). Microrheology with optical tweezers: Measuring the relative viscosity of solutions at a glance. *Scientific Reports, 5*, 8831.

85. Vadillo, D. C., Tuladhar, T. R., Mulji, A. C., Mackley, M. R., Jung, S., & Hoath, S. D. (2010). The development of the 'Cambridge Trimaster' filament stretch and break-up device for the evaluation of ink jet fluids. *Journal of Rheology, 54*(2), 261–282.

86. Vandebril, S., Franck, A., Fuller, G. G., Moldenaers, P., & Vermant, J. (2010). A double wall-ring geometry for interfacial shear rheometry. *Rheologica Acta, 49*(2), 131–144.

87. Verwijlen, T., Leiske, D., Moldenaers, P., Vermant, J., & Fuller, G. G. (2012). Extensional rheometry at interfaces: Analysis of the Cambridge interfacial tensiometer. *Journal of Rheology, 56*(5), 1225.

88. Waigh, T. A. (2005). Microrheology of complex fluids. *Reports on Progress in Physics, 68*(3), 685.

89. Zimmerman, W. B., Rees, J. M., & Craven, T. J. (2006). The rheometry of non-Newtonian electrokinetic flow in a microchannel T-junction. *Microfluidics and Nanofluidics, 2*, 481–492.

90. Zimmerman, W. C., & Rees, J. M. (2013). Rhemeter and rheometric method. WO Patent App. PCT/GB2013/051,089.

91. Zografos, K., Pimenta, F., Alves, M. A., & Oliveira, M. S. A. (2016). Microfluidic converging/diverging channels optimised for homogeneous extensional deformation. *Biomicrofluidics, 10*, 043508.

Chapter 2
Microfabrication Techniques for Microfluidic Devices

Vania Silverio and Susana Cardoso de Freitas

2.1 Introduction

Microfluidic technology usually, but not exclusively, aims at the miniaturization of conventional laboratorial processes to handle fluids (liquids and/or gases) within submillimeter ranges. The manipulation of fluids follows different strategies (e.g., sample/reagent actuation, mixing, separation, filtration, reaction, control, monitoring, detection, etc.) depending on the final objective and available space for operation. Recent efforts focus on integration of e.g., micro-to-nano elements, sensing, electronics or automation for high throughput and low-cost devices [3]. Additionally, microfabrication strategies aim at providing unique and increased functionality for sample-to-answer compact platforms for intelligent systems design towards e.g., biological, medical or chemical applications.

The fabrication methods and their limitations together with the application largely determine the size, shape and material of microchannels and the layout configuration of their arrays [29]. The correct selection of material used in the microfabrication process is necessary for the successful application of advanced and low-cost microfluidic devices. Common viable compatible materials used are silicon, metal, silica, polymeric or paper [28].

Fabrication techniques used for `silicon-based` microfluidic devices are mostly based in Micro-Electro-Mechanical Systems (MEMS) methods, which have dramatically developed alongside semiconductor technology [38]. MEMS fabrication adopted techniques well established in the integrated circuit (IC) industry. These include oxidation, ion implantation, low pressure chemical vapor deposition

V. Silverio (✉) · S. Cardoso de Freitas
INESC Microsystems and Nanotechnologies, Rua Alves Redol, 9,
1000-029 Lisboa, Portugal
e-mail: vsilverio@inesc-mn.pt

S. Cardoso de Freitas
e-mail: scardoso@inesc-mn.pt

© Springer International Publishing AG 2018
F.J. Galindo-Rosales (ed.), *Complex Fluid-Flows in Microfluidics*,
DOI 10.1007/978-3-319-59593-1_2

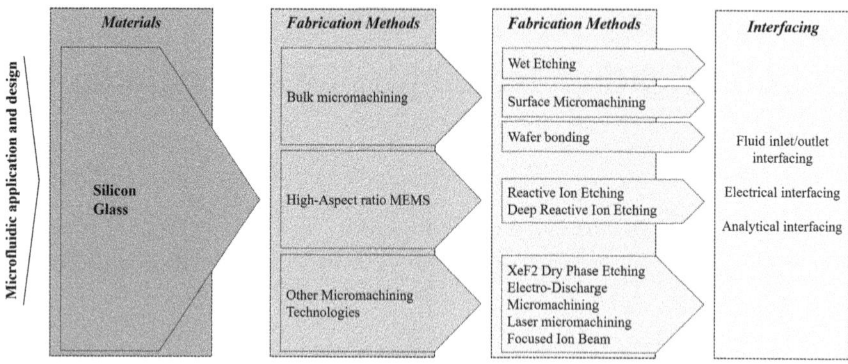

Fig. 2.1 Fabrication methods for rigid substrates

(LPCVD), diffusion, sputtering, etc. Actuation, control and measurement functions can be simultaneously microfabricated on silicon wafers by combining IC techniques with dedicated micromachining processes e.g., deposition of SiO_2 as a sacrificial layer and for electrical insulation, polysilicon deposition for conductive or resistive features, metal deposition for conductive elements or TiN and Si_3N_4 deposition for passivation and electrical insulation (Fig. 2.1).

Depending on the type of microfabricated systems, they can go from 20 nm to 50 μm feature size. Recording heads or MRAM (Magnetoresistive Random-Access Memory) technologies for example, must have features below 100 nm to be competitive [40] whereas dimensions of e.g., RF and power transistors are set from 100 nm to 1 μm [12]. MEMS devices typically have 1–10 μm minimum features and microuidic devices might have features as small as 5–10 μm, but are usually larger.

On the other hand, using `polymer materials` to fabricate microfluidic devices (Fig. 2.2) reduces fabrication times while providing simple, cost effective, and disposal advantages when compared to Silicon. Polydimethylsiloxane (PDMS) elastomer and thermoplastics (polyethylene terephthalate—PET, polymethyl methacrylate—PMMA, polycarbonate—PC, or polyimide—PI) are the major polymeric materials used in microfuidics mainly due to their flexibility, optical transparency to visible/UV, easiness of molding with high fidelity, simplicity in the modification of surface properties, biocompatibility or bioactivity, durability, chemical inertness and low-toxicity. Contrarily, low thermal stability and low thermal and electrical conductivity can severely limit the choice of polymeric materials for microfluidic applications.

Rapid prototyping of the microfluidic devices in `elastomers` is achieved using soft lithography, where the PDMS is poured over a master mold to create the required fluidic passages. The molding master is fabricated by either patterning positive photoresist, e.g., AZ 40XT or negative photoresist, e.g., SU-8, on a silicon substrate for precise microstructures (sizes below 50 μm) or by direct micromilling of polymeric material (e.g., PMMA) for larger microstructures (sizes above 50 μm). These process are simple and inexpensive, yet time consuming and unviable for mass production.

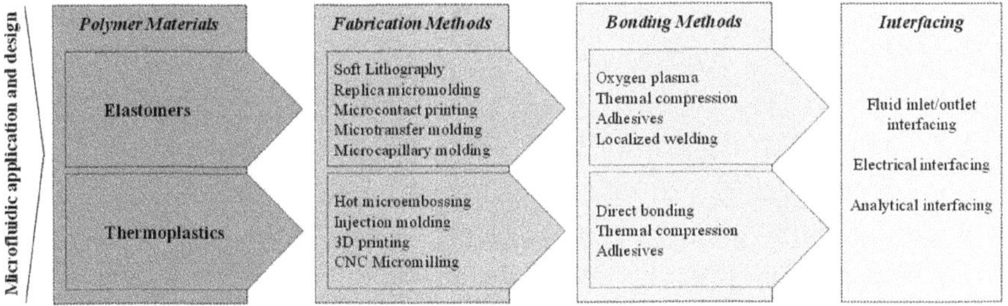

Fig. 2.2 Microfabrication techniques for polymer based microfluidic devices

Alternatively, `thermoplastics` are softened or fully melted and reshaped upon heating. Various methods exist for the fabrication of thermoplastic microdevices e.g., hot embossing, injection molding, micromilling or 3D printing. The process is implemented guaranteeing thermoplastics remain chemically and dimensionally stable over the range of operational temperatures and pressures [43]. Depending on the fabrication process and molds, devices up to a few micrometers can be fabricated onto thermoplastics to high impact resistances. From the fabrication processes presented, hot embossing is viewed as the most commercially viable for mass production [16].

`Paper microfluidics` has been recently proposed as a convenient alternative for a new class of microfluidic devices where the manipulation of fluids is accomplished within paper-like porous materials [8]. Paper is a low-priced, lightweighted and disposable material, ready accessible in a wide spectrum of thicknesses, making it extremely attractive for versatile, low-cost, easy to use and fast sample-to-answer microfluidic devices [31]. However, small and imprecise sample volumes limit precision and sensitivity of paper microfluidic devices. Moreover, their design aims at individual testing, making them unsuitable for high-throughput screening.

`Bonding` is a critical step in microfabrication processes necessary to provide a tight seal to prevent leakage without deformation of the channels. Bonding can be chemical (e.g., UV/O_3 or plasma activation), thermal (compression at temperature elevated to the glass transition T_{glass}), or adhesive (e.g., resins or curable materials), etc. [42]. A large variety of materials can be joined to form microfluidic passages which may include sensing and actuation. Two distinct materials can be bonded together (e.g., silicon/glass, glass/polymer, etc.) or the channels can be fabricated with several parts of the same material.

The choice of the fabrication method relies on its capability to create the part with the desired features. Other technical capabilities have to do with the compatibility with the material to use and the quality of the finished part. With this in mind, microfabrication techniques for microfluidic devices (examples in Fig. 2.3) are detailed below. The discussion of a set of unique challenges such as the best material and adequate fabrication process to be used for particular applications and the total cost of fabrication is also presented.

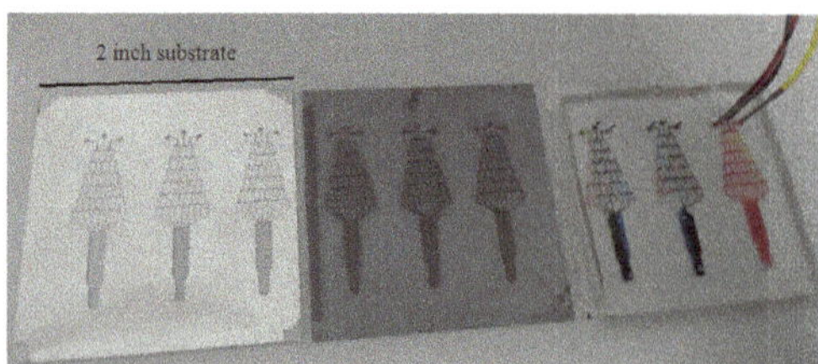

(a) Hardmask, mold and PDMS microfluidic device

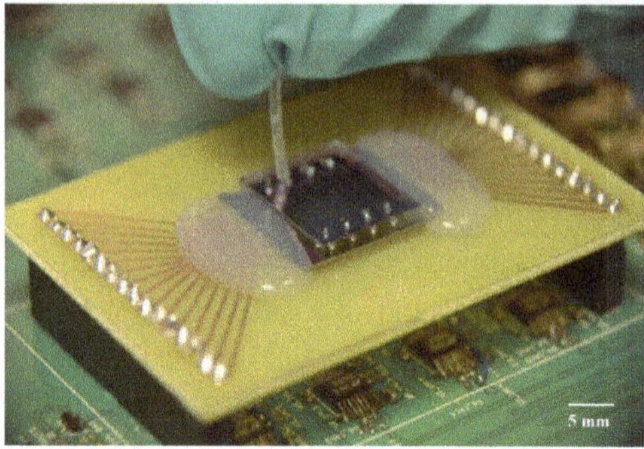

(b) Integration of PDMS microfluidic device
with magnetic sensors and electronics

(c) PMMA microfluidic device
fabricated by 3D micromilling

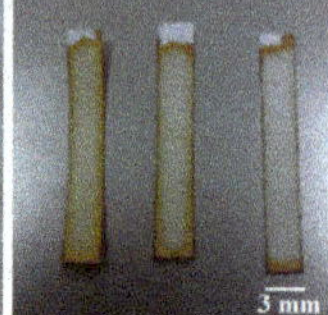

(d) Microfluidics on paper

Fig. 2.3 Examples of microfluidic devices

2.2 Fabrication of Microfluidic Devices in Rigid Substrates: Silicon/Glass

Microfabrication of microfluidic devices in rigid substrates makes use of well established techniques of semiconductor industry such as lithography, subtractive and/or additive techniques (e.g., etching and thin-film deposition), to achieve few to hundreds of micrometer devices on top or within the substrate. Microfabrication techniques can be used sole or combined to obtain simple or intricate devices as reported below.

2.2.1 Lithography

Lithography is the process used to transfer three-dimensional patterns onto a surface [32]. This is seen as the most important step in the microfabrication process as the definition of the shape of elements constituting the devices is performed here.

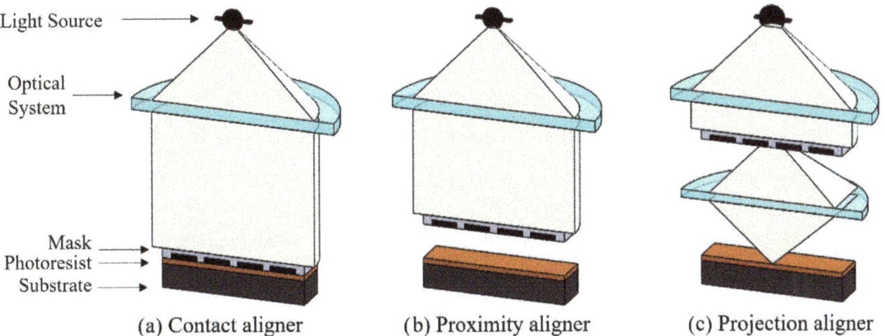

Fig. 2.4 Methods to transfer a pattern to a surface

Lithography steps: (1) designing the pattern, (2) making the mask, (3) coating the wafer, (4) exposing the photoresist, (5) developing the photoresist.

Several steps define the lithographic process. The geometric pattern is designed using computer-aided design (CAD). The pattern is then transferred to a light-sensitive polymer, (also called photoresist, PR) previously coated on the substrate, by exposing the PR to light (UV, Laser, beam of electrons). The light can pass through a hardmask or a photomask (photolithography) to expose the PR or the lithography can be maskless, as is the case of Laser lithography. After the development of the PR, the substrate is ready to proceed for the next step of fabrication.

The PR is exposed to light under programmed duration and energy, determined by its energy adsorption. `Laser lithography` is the lithographic technique using laser light to fabricate large precision patterns. The minimum feature size achieved with this technique ($\sim$0.8 μm or less) depends on the laser light wavelength and optical path. Additionally, modulating the intensity of the laser can result in partially exposed PR (Laser Direct write grayscale photolithography), which can be used to create microscale features with multilevel topography. `Photolithography`, also called optical or UV lithography, predominantly uses UV light through a photomask or hardmask to expose the PR. The photomask or hardmask is usually defined on a metallic thin film layer (e.g., aluminum, chromium, etc.) opaque to UV light, deposited onto a flat UV-transparent substrate (Table 2.1). The geometric patterns are usually determined by computer-aided design for increased definition and can be defined over large sized substrates (up to $\sim$30 cm).

The strategy of exposure will define the process resolution, which is diffraction limited. For example, in `contact lithography` (Fig. 2.4a), the hardmask is in direct contact with the PR. The features to be transferred are so at a 1:1 ratio (no magnification) for `minimum feature sizes` (m.f.s.) of 0.5 μm. Although this is a very simple to use and inexpensive technique, the direct contact with the mask usually smears and degrades the features, leading to a loss of sharpness and

Table 2.1 Types of basic materials used to make photolithography masks

Quartz	Soda lime	Plastic mask
Expensive	Good price/quality ratio	Low price
High resolution	High resolution	Low resolution
Very stable	Stable	Weak stability
Can break	Can break	Easy to handle
Wavelength >180 nm	Wavelength >350 nm	Wavelength >350 nm

definition. Exposure strategies where the mask is not in direct contact with the PR, are either `proximity printing` (small gap between mask and PR, magnification 1:1) Fig. 2.4b or `projection printing` (projection from a distance and demagnification 1:4–1:10 or even lower) Fig. 2.4c. In proximity printing, the diffraction effects limit the accuracy and repeatability of pattern transfer and resolution worsens ($\sim$1–2 μm). The optical path in projection printing defines the resolution of the process, which can achieve m.f.s. of $\sim$0.065 μm or slightly better, however, it is also the most expensive of the three.

`Photoresist` is a light-sensitive polymeric solution. Negative PR (the regions exposed to light become insoluble in the developer) originate m.f.s. up to 2 μm and spin-coated spreading thickness from 1 to 3 μm (the photoresist film thickness is determined by the spinning speed), while positive PR (the regions exposed to light become soluble in the developer) has m.f.s. of $\sim$1 μm for spreading thicknesses between 1.5 to 7 μm.

The m.f.s. on the pattern are also defined by the type of PR used in the process. After exposure, the development of PR is pursued to remove the unwanted regions. The resolution and definition of fabricated patterns is directly dependent of the development time and conditions (e.g., temperature, agitation, concentration, etc.). The PR still remaining on the substrate will define etching regions or deposition regions either remove or add material in a subsequent fabrication step.

2.2.2 Subtractive Techniques for Pattern Transfer Microfabrication

The etching process is a subtractive process in which the pattern is transferred by chemical/physical removal of the material. Generally two approaches are followed depending on the etching strategy. Wet etching uses chemical solutions as dry etching is performed through plasma methods. Additionally, a combination of the two methods (reactive ion etching) if often used.

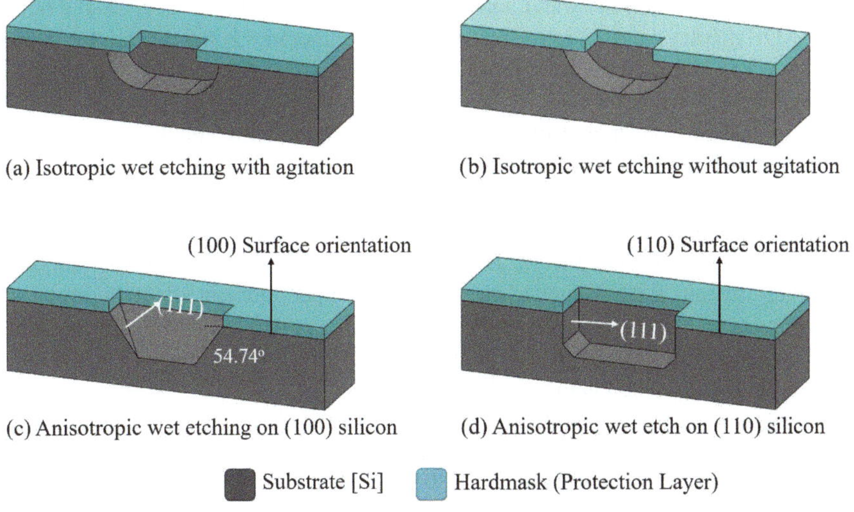

Fig. 2.5 Characteristic etched profiles obtained by isotropic or anisotropic etching

2.2.2.1 Wet Etching

Wet etching involves using liquid etchant to remove material from a surface. This is achieved isotropically or anisotropically (Fig. 2.5). Anisotropic etching uses crystal-orientation etchants of silicon (e.g., hydroxides of alkali metals: RbOH, KOH, CeOH, NaOH, ethlenediamine-pyrocatechol solution-EDP process). Planes (111)[1] are etched slower than other planes. Moreover, etching at concave corners on Si (100) planes stop at (111) intersections creating walls with 54.74° angle to the surface while convex corners are undercut.

- Isotropic etch : surface material is removed uniformly in all directions of the chemical structure.
- Anisotropic etch: the removal of surface material is dependent on the crystalline structure orientation of the surface.

Although very important in a fluid dynamics point of view, the EDP process is incompatible with MOS (Metal Oxide Semiconductor) or CMOS (Complementary Metal Oxide Semiconductor) processing as it is highly corrosive consequently rusting

---

[1] The orientation of a crystal plane or a surface are defined considering the plane intersection with the main crystallographic axes of the solid. The lattice directions and planes are mathematically described by the Miller Indices. The Miller Index notation is used as follows: family of directions $< \cdots >$, particular direction $[\cdots]$, family of planes $\{ \cdots \}$, particular plane $(\cdots)$.

any metal deposited on the wafer and it permanently stains the surface. On the other hand, KOH displays an etch rate selectivity 400× higher in <100> crystal directions than in <111> directions making it extremely suitable for selective etching, as opposed to EDP whose selectivity for <100> is only 17× higher than for <111>, mobile potassium ions may drift to silicon dioxide or violently attack exposed metal contacts if etch rates higher than 1 μm.min^{-1} are used [27].

Wet isotropic etching (orientation independent) makes use of aqueous acidic solutions containing HF (hydrogen fluoride) and HNO (azanone) or HNA (HNA is a mixture of HF, CH_3COOH (acetic acid) and HNO_3 (nitric acid)) to remove materials from the substrate. Factors such as agitation and temperature can affect the time and homogeneity of the process. The final etched shape is defined by the composition of the surface [35]. Characteristic round corner structures are obtained by isotropic etching of silicon, which is limited by diffusion.

> The etch rate in wet etching can be enhanced by tuning agitation and temperature.

Fig. 2.6 Physical dry etching, chemical dry etching and reactive ion etching mechanisms

Physical Dry Etching

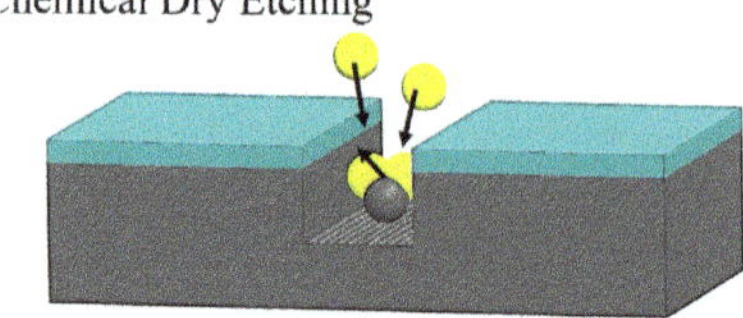

Plasma atoms hit the surface and silicon atoms are evaporated from the surface

Chemical Dry Etching

Interaction between reactive ion and silicon atom; bond between reactive ion and silicon atom; chemical removal of silicon atoms from the surface

Reactive Ion Etching

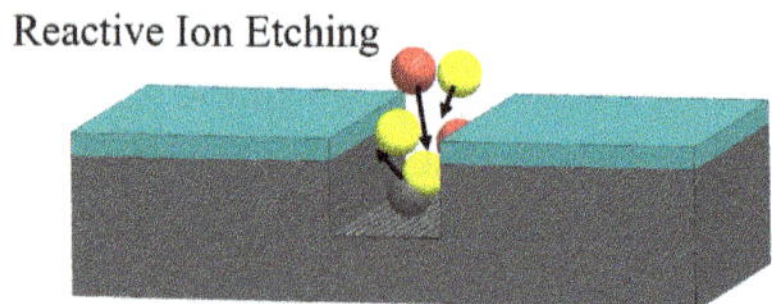

Physical and chemical reactions to etch the silicon

Substrate [Si]   Mask   Plasma   Silicon atom   Reactive ion

Both anisotropic and isotropic wet etchants can produce only a few specific channel geometries, limited by the wafer type and channel orientation, and lead to largely biased dimensions as well. Nonetheless, wet etching is a fast and simple etching technique suitable for low cost production.

2.2.2.2 Dry Etching

Dry etching uses plasma or etchant gases to remove material from the substrate at low pressure (few millitorr) and low temperature (from room temperature to 250 °C). The reactions taking place use high kinetic energy of particle beams (ion, electron or photon) to etch the substrate atoms (physical dry etching), chemical reactions between etchant gases to attack the surface (chemical dry etching) or a combination of both (reactive ion etching—RIE) to achieve high resolution (Fig. 2.6). Chemical dry etching uses reactive gases such as CH_4, SF_6, NF_3, Cl_2 and F_2. RIE is one of the most used in research and industrial processes. Typical RIE gases for silicon substrates are CF_4, SF_6 and $BCl_2 + Cl_2$. The reactions proceed almost spontaneously originating nearly isotropic profiles and, therefore, very precise pattern transfers.

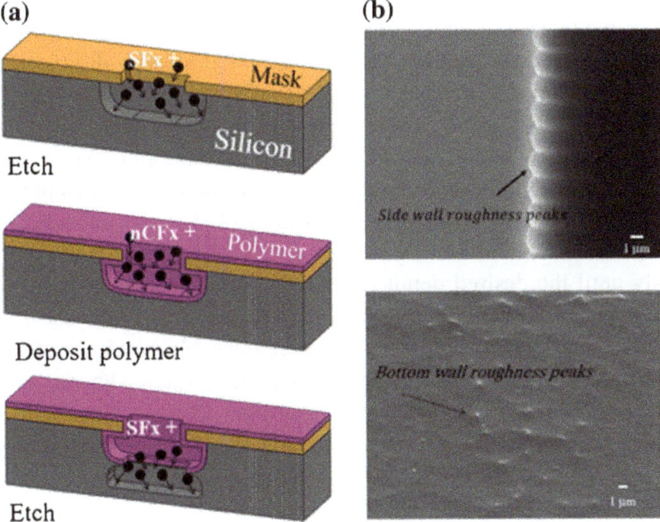

Fig. 2.7 a Schematics of the DRIE process; **b** SEM of microchannel side wall roughness peaks (M 5000×) and SEM of roughness peaks on the microchannel bottom wall (M 3300×), resulting from the Bosh process

2.2.2.3 High-aspect-ratio: Deep Reactive Ion Etching

`Deep Reactive Ion Etching` (DRIE) is a relatively recent fabrication technique which enables high aspect ratio elements to be defined into silicon substrates, as well as into metal or glass.

> The DRIE process provides deep features of hundreds to thousands of micrometers into the substrate with nearly vertical walls.

The Bosch etch is one of the most commercially employed DRIE technique [25], which uses alternating etch/passivation/etch chemistries. The technique is known to form these deep vertical walls with a characteristic scalloped sidewall tipically with a peak-to-peak roughness of about 0.3 μm [17]. It requires laser lithography to outline the patterns in the photoresist mask which will define the unprotected areas of the substrate to be etched.

Although time consuming and expensive, the DRIE fabrication technique has very low manufacturing uncertainty. Moreover, this method can potentially manufacture an unlimited number of geometries and achieve feature sizes of the order of a few micrometers, which makes it very attractive for the growing field of integrated sensors and actuators and microfabrication of silicon microchannels.

The DRIE process is as follows: The etch uses high-density dry plasma to alternately etch the silicon and deposit a layer of polymer resistant to etching on the side walls. The process is done as many times as to achieve deep, vertical etch profiles. Silicon etching is conducted using sulphur hexafluoride (SF_6) chemistry to provide the free radical fluorine in high-density plasma for silicon etching Fig. 2.7a. The protection of the side walls and the bottom of the etch pit is accomplished with the deposition of octofluorocyclobutane (c-C_4F_8). These two steps alternate continuously until the desired depth is attained and depend largely on the size of the silicon area exposed to etching. Large unprotected areas of the substrate etch at much larger etch rates when compared to smaller areas. Moreover, the etch rate at the bottom pit is higher than that at the walls and a characteristic washboard or scalloping pattern is seen in the side walls due to the anisotropic nature of the process Fig. 2.7b. The etch rates on most commercial DRIE systems varies from 1 to 4 μm.min^{-1}.

2.2.3 Additive Techniques for Pattern Transfer

Additive microfabrication techniques are based on the addition of materials, usually selectively, to a substrate (Fig. 2.8). The addition can be mechanical, chemical or thermal depending on the energy used for adhesion.

Physical vapor deposition, PVD, is one of the methods used in the fabrication of microfluidic devices. It consists in depositing thin films onto the substrate by condensable vapor of the material to deposit through low-pressure or vacuum gaseous

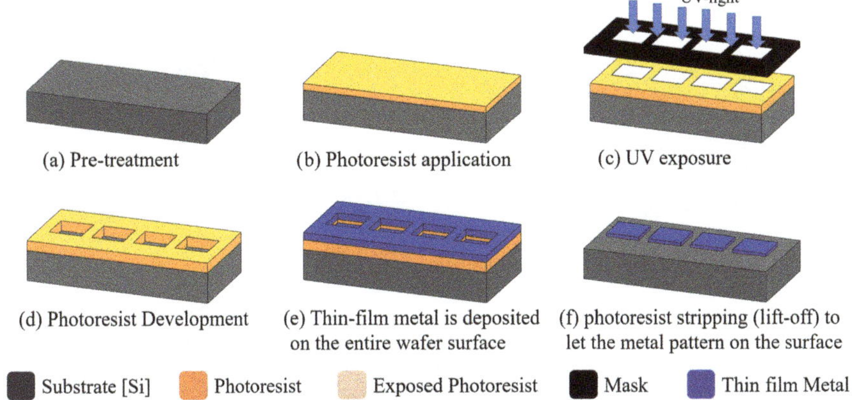

Fig. 2.8 Common fabrication steps for patterning metal onto a silicon wafer

environment. The vapor is shifted from the target source to the rigid substrate by various techniques (e.g., sputtering, thermal evaporation, ion plating and cluster deposition, laser ablation or laser sputter deposition, aerosol deposition, etc.).

Additive microfabrication steps: (1) synthesize the vapor coating from the source (target) material, (2) transport the vapor to the substrate, (3) condense the vapor coating onto the surface of the substrate.

Sputtering deposition consists in bombarding a sputtering target by accelerated inert ions (e.g., Ar^+) in plasma atmosphere. The ions impinging on the target erode (sputter etch) the target surface by momentum transfer. The plasma can be obtained by DC (direct current), reactive sputtering, radio frequency (RF), magnetron, etc. The material ejected from the target becomes condensable vapor for deposition.

Thermal evaporation is used to deposit mainly metals (e.g., Au, Al, Ti, Cr) or compounds with low fusion temperature (e.g., SiO_2, Si_3N_4) by means of resistive or electron beam heating. Unlike sputtering, evaporation deposits directionally from the source and sidewall coating is poor when compared to surface coating. Additionally, this technique provides lower deposition rates at higher substrate temperature than sputtering, generating thin films with higher tensile stresses.

Surface micromachining is characterized by the fabrication of micromechanical structures from thin films deposited above the surface of the silicon substrate [2]. Thin-film deposition is an additive process where layers of thin films (e.g., silicon dioxide, polysilicon, silicon nitride, and metal) are added to the substrate surface.

An example of the combination of surface micromachining and wet etching for the fabrication of a cantilever is shown in Fig. 2.9. The process starts with the deposition

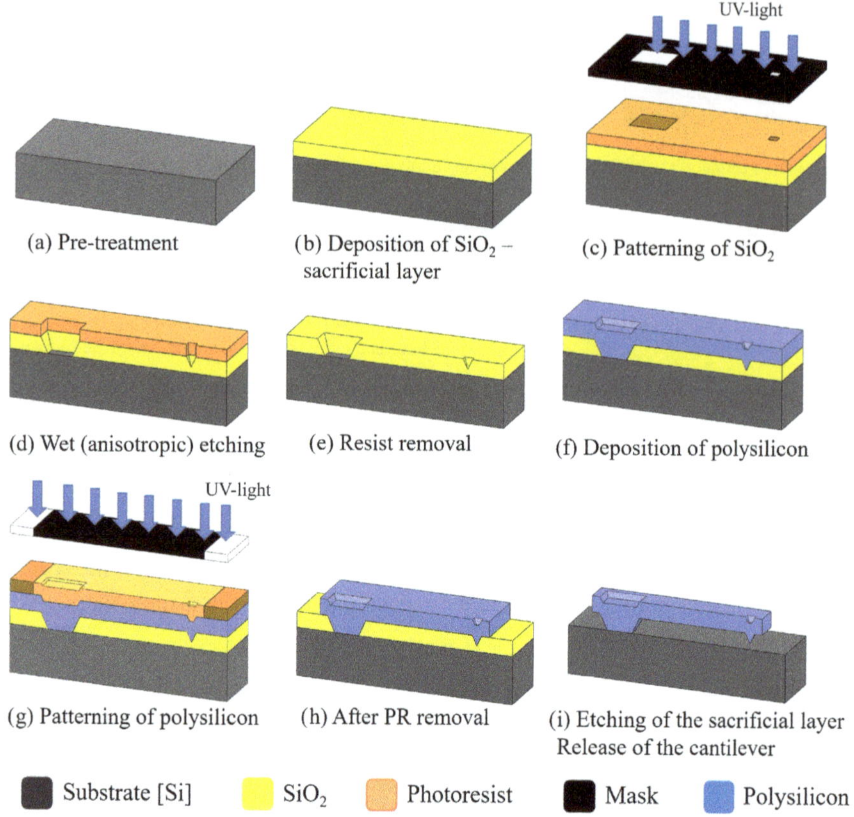

Fig. 2.9 Surface micromachining as a fabrication process for a cantilever

of a thin film material to act as a temporary mechanical layer also called sacrificial layer, onto which the subsequent device layers are built (Fig. 2.9b); next, patterning and etching of this sacrificial layer the deposition and patterning of the structural layer of thin film material (Fig. 2.9f, g). Etching of the sacrificial layer and release of the cantilever finalize the process (Fig. 2.9i), allowing the polysilicon structural mechanical layer to move.

2.3 Microfabrication Techniques for Polymer Based Microfluidic Devices

Along with silicon and glass microfabrication, cheaper and easier solutions are being followed, particularly in biological detection applications. Polymeric materials are gaining considerable attention as raw materials for microfabrication of fluidic passages. They are easily replicable, transparent, biocompatible and have thermal and electrical properties suitable for e.g., biological detection applications. Two main types of polymeric materials (elastomers and thermoplastics) and their microfabrication techniques are addressed below.

2.3.1 Elastomers: PDMS

PDMS has become the most commonly used elastomer in rapid prototyping of microfluidic devices because of its simple fabrication procedure through casting and curing onto a microscale mold and strong sealing to a wide variety of materials.

> The PDMS is composed of a two-part heat-curable mixture. The pre-polymer is cross-linked with the curing agent usually in a 10:1 ratio in weight, but varying this ratio one can obtain different mechanical and chemical properties of the resulting mixture.

PDMS microfluidic fabrication gains from PDMS low cost, biocompatibility, chemical inertness, low toxicity, versatility on surface chemistry insulating, as well as mechanical flexibility, ease of manipulation and durability. The low stiffness of PDMS ($\sim$1 MPa) has additionally been explored for the integration of valves and pumps and to produce 3D multilayered devices [1, 7, 39]. But difficulties in process repeatability and handling such a soft material frustrates fabrication scale-up. PDMS is very sensitive to minute changes in applied pressures. As a result, compression of PDMS structures may lead to cracks and deformation of the feature dimensions. Furthermore, PDMS is more suitable for low aspect-ratio channels as high aspect-ratio channels are difficult to obtain and are prone to collapse.

Gas and water permeability of PDMS and its hydrophobic nature work for and against its usage. For example, non-specific protein/hydrophobic analyte can be absorbed by the PDMS due to strong hydrophobic-hydrophobic interactions [48]. Cell or bacteria can adhere to the PDMS walls, [11]. While in some applications this is the desirable goal [14], if ill designed, such bioanalytes can obstruct the flow and ultimately lead to device fouling.

> `Soft lithography` is the technique used to rapidly and easily fabricate and replicate a wide range of elastomeric devices, i.e., mechanically soft materials (e.g., polymers, gels or organic monolayers) without costly capital equipment.

Soft lithography can be viewed as a prolongation of photolithography. After the definition of the mask (Sect. 4.2.1), the master mold to be used for the definition

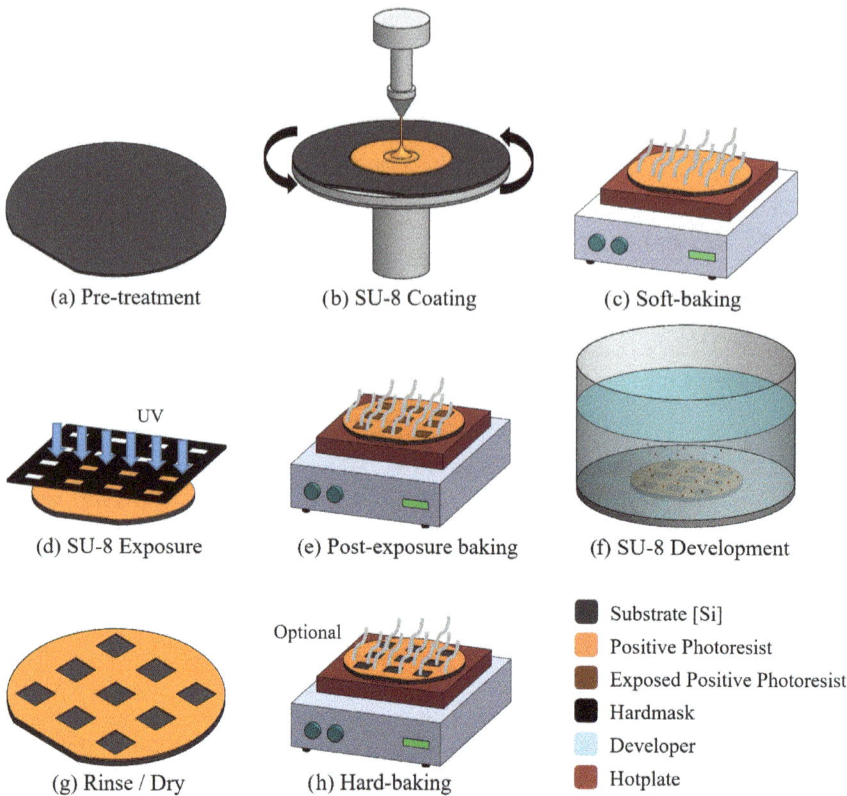

Fig. 2.10 Fabrication of the master mold by photolithography

of elastomeric structures needs to be fabricated. The molding template is fabricated by either patterning photoresist on a flat rigid substrate (e.g., silicon wafer) for precise microstructures (e.g., $<50\,\mu m$) or by micromilling thermoplastic polymer substrates for larger microstructures (e.g., $>50\,\mu m$). A brief description of photoresist patterning is given below. Thermoplastic microfabrication will be referred in detail in Sect. 4.3.2.

The `master mold` fabricated by photolithography onto PR roughly follows the steps presented in Fig. 2.10. The photoresist manufacturer's datasheet presents the exact steps and specification/tolerances to be used in the photolithographic process. The process starts with a thorough cleaning of the substrate (Fig. 2.10a). Harsh cleaning methods include piranha clean SOP, reactive ion etching or plasma ashing, but softer cleaning with an anionic detergent e.g., alconox, followed by solvents such as isopropyl alcohol is usually effective. De-ionized water rinse is then followed by drying with a nitrogen or compressed air stream.

To prevent any entrapped humidity at the surface, the substrate needs to be dehydrated through a thermal treatment (e.g., 200 °C for 5 min). The PR (e.g., SU-8 negative PR, AZ 40XT positive PR, etc.) is then dispensed onto the substrate and spread evenly using spinning (Fig. 2.10b). The speed, acceleration and time of spinning are conditioned by PR properties (temperature dependent) and will be defined according to the intended mold thickness. The maximum aspect ratio (width-to-height ratio) of the elements to be fabricated in the mold are dependent on the PR characteristics. Nevertheless, the aspect ratios in the final molded structure e.g., in PDMS, need to be designed guaranteeing the usability and robustness of the microfluidic device.

After coating, a soft-bake step is performed to evaporate PR solvents and prevent shrinkage and cracking of the PR in subsequent steps (Fig. 2.10c). The soft-bake process temperature and time are dependent on the PR formulation and mold thickness. Exposure to UV light through the mask defines the shape of the elements in the mold (Fig. 2.10d). The correct exposure energy and duration are strongly dependent on PR formulation, PR thickness and substrate material.

The definition of PR elements derives from its change in chemical composition due to exposure to UV light. Negative PR exposed regions become insoluble in the developer solution while the unexposed regions remain soluble. The opposite occurs for positive PR.

A post-exposure bake step is conducted to further harden the insoluble regions of PR making it resistant to the action of the developer solution (Fig. 2.10e). Again, the baking procedure is dependent on the PR formulation and mold thickness. The substrate is slowly cooled to prevent the formation of cracks due to thermal stresses. The substrate is now ready for the development step (Fig. 2.10f). The developer solution is PR specific.

The development conditions are dependent on the PR formulation and mold thickness. The development times used not only depend on the PR dissolution rates but also on the agitation conducted to promote dissolution. The completeness of development step is confirmed after washing the substrate with isopropyl alcohol and blow drying. The presence of white precipitates on the structures confirm underdevelopment, so more development time may be required for process completion. The mold is ready for the next step of fabrication after rinse with de-ionized water and blow drying (Fig. 2.10g). An extra step for hard-baking may be required for applications where the imaged resist is to be left as part of the final device.

The PDMS is casted onto the mold and cured to increase bond strength. The curing time and temperature define the structure mechanical properties. After curing, the elastomer is peeled from the mold and is now ready to be bonded to a rigid or flexible component to form the microfluidic device or to be used as a microstructured master.

Core techniques to fabricate patterns and structures based on the use of a patterned layer of elastomer are e.g., replica micromolding, microcontact printing (μCP), microtransfer molding (μTM), and microcapillary molding (MIMIC) (Fig. 2.11).

`Replica micromolding` is used to replicate soft microstructured masters on polymers. The polymer (e.g., agar, agarose, other biocompatible polymers) is poured on the soft mold and cured, after which is separated. Similarly to a rigid master used in photolithography (e.g., photoresist), the 3D elements defined onto the soft mold are imprinted onto the surface of the polymer. The soft mold can be reused several times before starting to collapse.

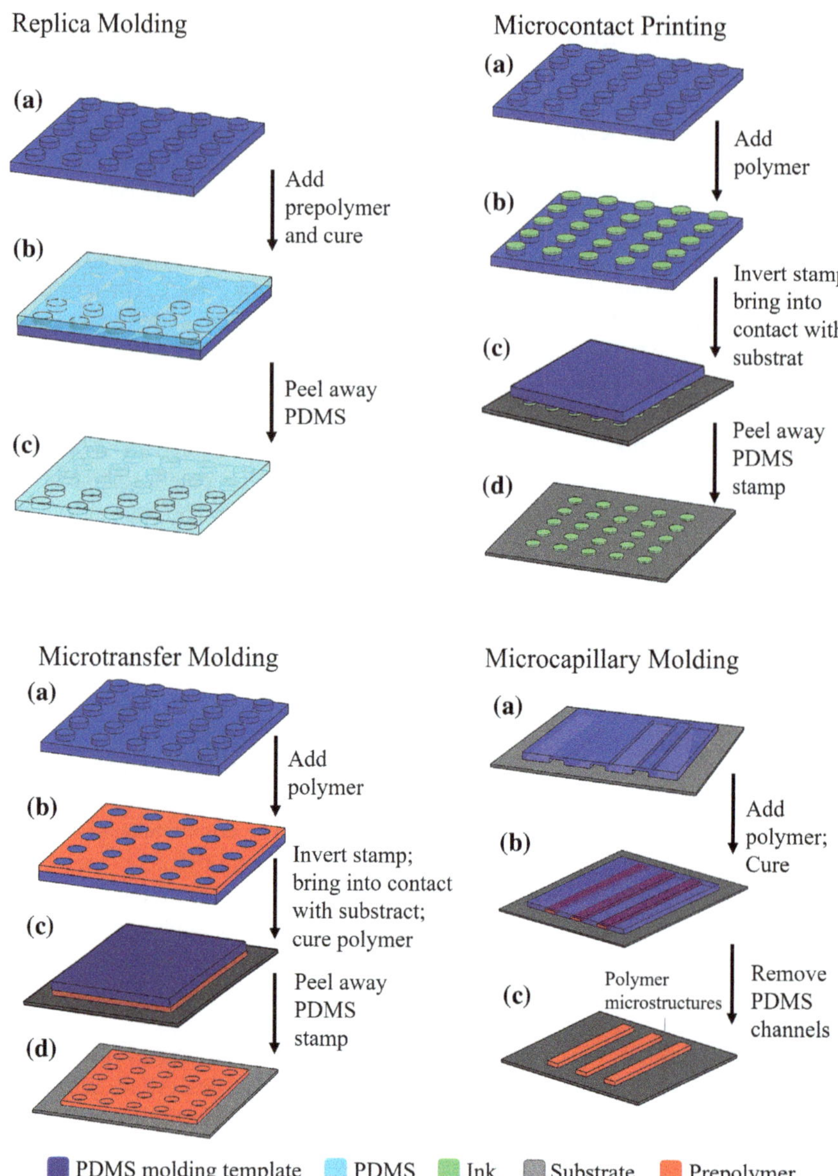

Fig. 2.11 Schematics of fabrication steps adopted in replica molding, microcontact printing, microtransfer molding and microcapillary molding

Microcontact printing (μCp) is a well-established soft-lithography technique consisting in transferring material (usually biomolecules or organic molecules) to the substrate in selectively-defined areas. The fabrication process, consists in four main steps. Firstly it is necessary to fabricate the PDMS stamp with the desired convex pattern using the soft-lithography steps described above. The PDMS stamp in then placed in contact with the material to be selectively transferred and subsequently pressed against the substrate surface to contain the material transfer. The stamp is then removed and the transferring material is retained at the substrate surface. μCp fabrication is relatively easy and cheap, and allows high-resolution patterns to be transferred into a wide variety of substrates e.g., glass [34], gold [6] or polymers [13] resulting in a diverseness of applications (e.g., microelectronics, cell biology or surface chemistry). The high compressibility of PDMS can be a major disadvantage [18], which may lead to ill-defined features and ultimately to the collapse of the stamp.

The micropatterned PDMS soft mold may be transferred to liquid polymer by `microtransfer molding` (μTM). The mold is filled with the liquid polymer which is turned, put in contact with the substrate and cured. The PDMS is then gently peeled from the polymer, leaving solid microstructures on top of the substrate. The PDMS soft mold can be reused similarly to replica molding. The PDMS soft mold can also define `microcapillary molded` (MIMIC) polymers. Here, the PDMS is brought in contact with the rigid substrate and the available volume between the two is filled with the liquid polymer by capillarity. Alternatively, the mold can be filled by suction. Curing, peeling and reusability steps are followed as defined previously.

2.3.2 Thermoplastic Polymers

Thermoplastic-based microfluidic devices are gaining attention mainly for their easiness towards large-scale production. Additionally from being cheaper in large-scale, thermoplastic fabrication allows faster processing with almost unlimited tailoring while not requiring clean-room environment. Thermoplastic properties such as high glass transition temperature, negligible impact resistance, low chemical absorption, biocompatibility, low intrinsic fluorescence and broad visible transmittance make them easy and appealing to use [24]. Of course, their chemical and physical properties can too be a drawback depending on the application. Some organic solvents may dissolve thermoplastics reducing their mechanical stiffness. Moreover, thermoplastics absorb UV-light inhibiting their use in a wide range of applications.

The possibility of rapid implementation of thermoplastic-based microfluidics allied to the necessity of reducing minimum feature sizes, increase bonding strengths, and fabricate intricate 3D solutions has been pushing investigation further. Techniques to fabricate microfluidic channels from thermoplastics include hot microembossing, injection micromolding, 3D printing, micromilling, LIGA and more recently, direct-write laser micromachining, among others [26]. Despite some of the polymeric-based techniques are still expensive and not readily available for microfluidic fabrication, they are slowly becoming more accessible as a result of intensified research efforts and growing interest in the field.

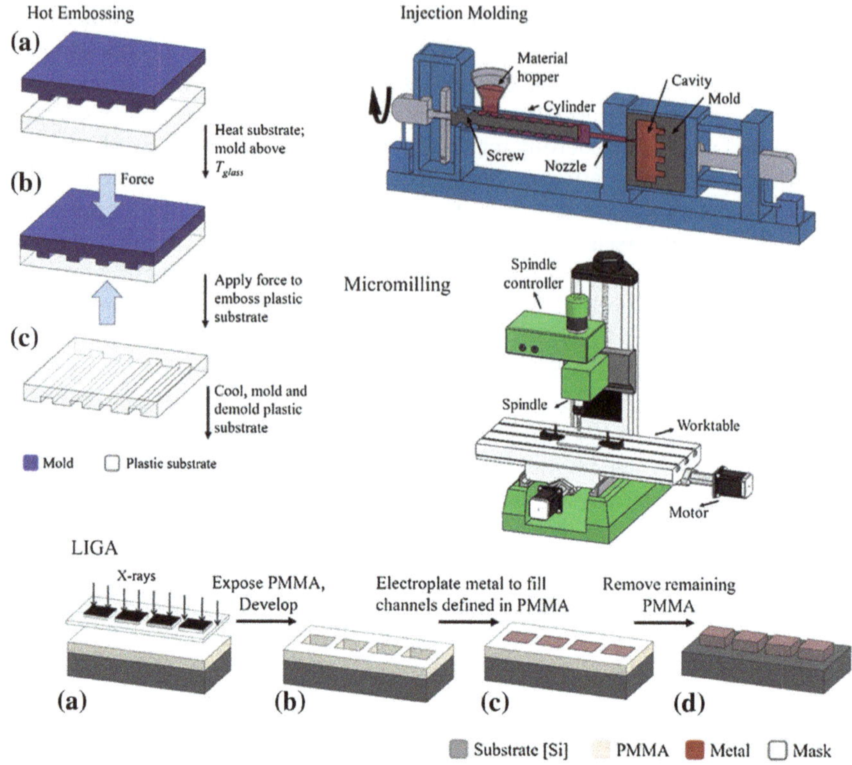

Fig. 2.12 Fabrication processes used to mold thermoplastic polymers

Micromolding the thermoplastics by `hot microembossing` consists in pressing the thermoplastic against the mold under controlled temperature and pressure (Fig. 2.12a). Operation temperatures are usually set above the glass transition temperature, T_{glass} of the thermoplastic [5], but never to a point where the thermoplastic is molten. Henceforth, high molecular weight polymers are well-suited for the process, provided their mechanical and thermal properties. The master mold can be PDMS [9], SU-8 [36] or other material, e.g., metals [22], which withstand molding operational conditions without deformation. After cooling, the thermoplastic polymer hardens.

Heating and cooling ramps define the mechanical properties of the molded part.

Hot microembossing of polymers which are transparent can increase the material autofluorescence, which can be important in fluorescence detection and microscopy [45]. More often than not, molded thermoplastics are too stiff to enable the integration of multilayers, valves and membranes. Additionally, high aspect ratio elements are difficult to fabricate and have large size dependent residual stresses. Nevertheless, large-scale production using this technique is straightforward for simple microfluidic structures [30].

Alternatively, `injection molding` includes steps for plastication, injection, packing, cooling and mold resetting. The polymer, usually pellets or granules, is plasticized melt and injected into the mold cavity containing the template for the features to be molded. As the polymer melt cools inside the mold, contraction occurs and is dependent on the polymer properties and application. A packing step is necessary to fill empty spaces caused by polymer shrinkage. The cooling step is defined by the cooling rate, which largely determines the success of fabrication. After this step, the molded polymeric piece is removed from the mold and the process is reset to step one.

This is a rather inexpensive microfabrication technique for large-scale production, capable of achieving high quality in a minimum amount of time. However, it is unsuitable for short production, as costs and time needed for the development and fabrication of custom molds can be excessive. Just as hot embossing, the aspect ratios, surface finishing, surface roughness and feature resolutions are not only dependent on the properties of materials used but also on the mold quality.

`3D printing` is an additive fabrication technique in which the polymer is heated while passing through the head of the printer. The piece is supported in a movable table enabling XYZ positioning. The passage of the printer head prints the polymer one layer at a time without requiring a mold.

While 3D printing is an additive process, `micromilling` is a subtractive fabrication process. It relies on the use of rotating cutting tools to remove bulk material from the workpiece. The system basically consists in a worktable for XY positioning of the workpiece, the cutting tool (usually endmill or drill) and a spindle for securing, Z positioning and rotating the cutting tool. For micron sized dimensions, the milling control is usually CNC (computer numerical control) for automation, repeatability and precision. This control enables direct conversion of CAD, reducing human error and leading to the fabrication of more intricate parts. Micromilling can be used to directly mill fluidic passages and features into the final device, but also to fabricate molds to be used in other fabrication techniques. At these scales, polymeric materials are the most used, contrarily to steel or copper used at larger scales. Elastomers can also be used, but due to their high elasticity, they are generally difficult to fabricate with this technique.

The `LIGA` process consists in having a mold, usually from PMMA (up to and exceeding 1 mm thick), which by previous exposure to X-ray (synchrotron) is developed to well defined, extremely smooth and nearly perfect vertical sidewalls. This technique can create structures $100\times$ deeper than wide, with submicron tolerances. For example, SU-8 can also be used as a mold for lower aspect ratio features. The mold is then placed in an electroplating bath and the filling metal (e.g., nickel, gold, copper, etc.) is electroplated from the conductive seed substrate into the open areas of the PMMA. The current density is set by controlling DC electroplating current or pulsed electroplating current. Advantages arise from a pulsed approach. The reactant

A. Technical Capabilities

Categories	Milling	Embossing	Stereolithography	Injection Molding
Material Capabilities				
Thermoplastics	●●●	●●●	●●○ *	●●●
Thermosets	●●○ *	●●○ *	●●○ *	●●●
Elastomers	●○○	●●○ †	●●○ *	●●●
Metals	●●●	●○○ ‡	●○○ †	●○○ *
Glass/Ceramics	●○○ †	●○○ §	●○○ †	○○○ *
Wax	●●●	●●●	○○○	●●●
Feature Capabilities				
Additional Heights	No added complexity	Additional layer per height	No added complexity	No added complexity
Aspect Ratio	8:1	2:1	Method dependent	8:1
Contoured 3D Features	Continuous	Layered	Layered	Continuous
Sharp Corners	External Only	Internal / External	Internal / External	Internal / External
Undercuts	Special tooling	Impractical	Yes	Special tooling
Results				
Surface Roughness	0.4 - 2 µm	Replicates mold roughness	0.4 - 6 µm	Replications mold roughness
Autofluorescence	Not affected	Increased by processing	Material dependent	Not affected

Legend

●●● Excellent
●●○ Most conditions
●○○ Specific conditions
○○○ Impractical

Milling:
* Only cured thermosets
† Poor consistency and characterization

Embossing:
* Only uncured thermosets
† Only thermoplastic elastomers
‡ Limited to specific features and thin sheets
§ Layered mix with polymer

Stereolithography:
* Uses resins, that when cured have similar properties to desired polymer
† Requires polymer/wax addative

Injection Molding:
* Requires polymer/wax addative

B. Cost Comparison

(Test Piece)

	Milling (On-site)		Milling (Outsourced)		Embossing (On-site)		Stereolithography (Outsourced)		Injection Molding (Outsourced)	
Setup Costs										
Equipment	$15k <		N/A		$15k <		N/A		N/A	
Tooling / Supplies	$500		N/A		$50		N/A		N/A	
Process Times and Costs										
	Time	Cost	Time	Cost	Time	Cost	Time	Cost	Time	Cost
Outsourced Expenses										
Mold / Tooling	N/A	N/A	N/A	N/A	4 - 15 d	$ 55 - 321	N/A	N/A	N/A	$ 2255
Device Fabrication	N/A	N/A	11 - 15 d	$ 137	N/A	N/A	4 - 6 d	$ 33	11 - 15 d	$ 2
On-site Expenses										
Machine Setup	10 m	N/A	N/A	N/A	5 m	N/A	N/A	N/A	N/A	N/A
Material Setup	< 5 m	$ 1	N/A	N/A	< 5 m	$ 1	N/A	N/A	N/A	N/A
Device Fabrication	10 m	N/A	N/A	N/A	30 m	N/A	N/A	N/A	N/A	N/A
Subtotal:	25 m	$ 1	N/A	N/A	40 m	$ 1	N/A	N/A	N/A	N/A
Expenses (per device)										
1 Devices	< 1 h	$ 1	11 - 15 d	$ 137	4 - 15 d	$ 56 - 322	4 - 6 d	$ 33	11 - 15 d	$ 2257
25 Devices	1 d	$ 1	11 - 15 d	$ 137	6 - 17 d	$ 3 - 14	4 - 6 d	$ 33	11 - 15 d	$ 92
50 Devices	3 d	$ 1	11 - 15 d	$ 137	8 - 19 d	$ 2 - 7	4 - 6 d	$ 33	11 - 15 d	$ 47

Fig. 2.13 Comparison of material, features, quality and cost of microfabrication processes used for plastics. Reprinted with permission from [10]. © 2015 Royal Society of Chemistry

species from the bulk electroplating solution are given time to reoccupy the region close to the cathode, where reactants are consumed in each electroplating pulse. This decreases plating stress and leads to more homogeneous and finer grain plating. The PMMA is stripped from the substrate to obtain the final structure which can be the final product or the replication mold used in the fabrication techniques previously described. The special mask and synchrotron radiation source make this process relatively expensive, henceforth, not widely used.

A comparison of the main material and feature capabilities is shown in Fig. 2.13.

2.4 Microfabrication Techniques for Paper Based Microfluidics

Paper-based microfluidics is gaining attention since its introduction in 2007 [44]. They are based on the fabrication of hydrophobic physical barriers on hydrophilic paper to define microfluidic passages where aqueous solutions will be passively transported. Fabrication methods reported are:

- photolithography—areas covered with negative PR (e.g., SU-8) become hydrophobic, while areas without photoresist continue hydrophilic [46]
- wax printing in a solid ink printer the heated wax melts and spreads vertically and laterally into the paper [4]
- PDMS dispensing—PDMS barriers are printed on filter paper after dissolved with hexanes [37]
- plasma treatment—the plasma treated areas are strongly hydrophilic [21].
  Tests on paper-based microfluidics are based on colorimetry, electrophoresis, electrochemistry or magnetic fields.

2.5 Bonding

Hybridization represents an important process to assemble the different parts of the device into a final product. Most microfluidic devices require bonding of several parts to produce confined volumes for fluid manipulation. A wide range of bonding strengths can be achieved depending on the materials and bonding technique [33].

> In general, the bonding techniques require flat, smooth and clean substrates to prevent voids between the bonded parts and achieve flawless bonding, even though they increase the costs of the overall fabrication and the material.

In addition, bonding may require a high temperature and high electrical field, which may damage the electronic elements on the device.

Silicon-to-silicon bonding can be employed by direct or fusion bonding techniques while anodic bonding is usually used to bond silicon-to-glass. Direct or fusion bonding (high-temperature annealing from 300 to 800 °C in oxygen or nitrogen atmosphere) is commonly used to bond two silicon substrates or a silicon substrate to a silicon oxide substrate, but other combinations can be pursued. For example, fusion bonding of bare silicon to a silicon substrate with a Si_3N_4 thin film deposit on the surface, or SOI-SOI (silicon on insulator) substrates.

Field assisted, anodic bonding or electrostatic sealing is a currently well-established industrial bonding technique that is reported to account for the majority of packaging applications for MEMS [23]. Its popularity is owed to high bonding yields, excellent capability in alignment and large process windows. It consists in a wafer bonding process to seal glass, containing a high concentration of alkali ions to either silicon or metal using an electric field and elevated temperature without introducing an intermediate layer. The conditions for bond surface quality are less restrictive than for direct or fusion bonding, substantially decreasing the process complexity.

The bonding process (Fig. 2.14a–c), of high bonding strength, consists in applying a DC voltage to the electrodes ensuring a positive electrode potential on the silicon side (Fig. 2.14a). The two substrates are placed in contact and compressed at $350-400°$ C (Fig. 2.14d). Thermal stresses are minimized if the thermal expansion coefficients of the two substrates are similar. Then, 900–1000 V direct current is constantly applied across the stack (glass substrate acts as the cathode). The sudden drop occurring in the applied current is due in part to the flow of Na^+ ions flowing to the cathode [19] (Fig. 2.14b). At $350-400°$ C, the alkali-metal ions in the glass migrate from the interface creating a depletion layer with high electric field strength and subsequent current flow of the oxygen anions from the glass to the silicon substrate results in an interface anodic reaction (Fig. 2.14c). As the migration of Na^+ continues, the concentration of positive charges repel incoming Na^+ ions inducing a

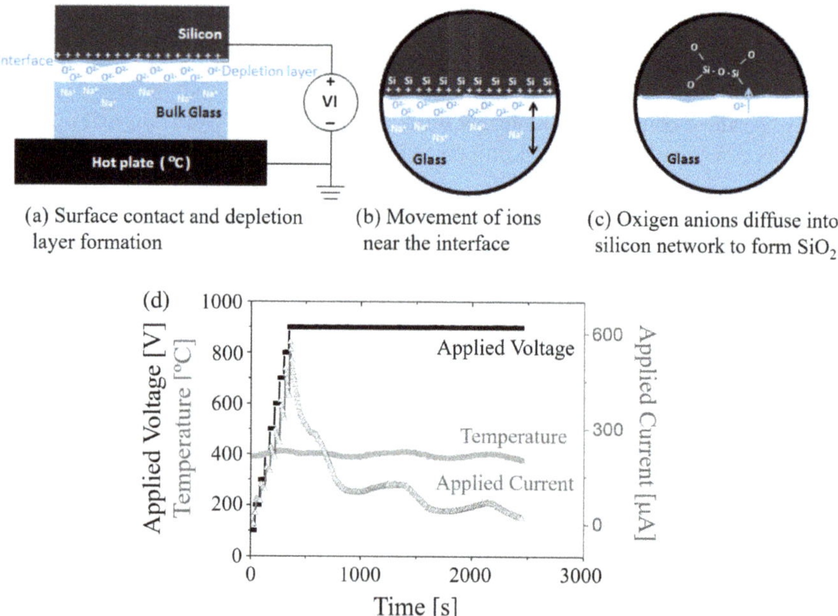

(a) Surface contact and depletion layer formation (b) Movement of ions near the interface (c) Oxigen anions diffuse into silicon network to form SiO_2

Fig. 2.14 Illustration of anodic bonding procedure. **a–c** Anodic bonding process. Pressure is applied to guarantee complete surface contact during ion diffusion and temperature is regulated to achieve the bonding layer formation, **d** experimental data of temperature, applied voltage and applied current during anodic bonding process

plateu in current which indicates bond completeness. The bonding seal between the two surfaces assures leakage contention without introducing dimensional errors to the microchannels.

Both fusion and anodic bonding are material specific bonding techniques providing great bonding strength. Generic hetero-bonding techniques such as `adhesive techniques` are the widelly applicable and substrate independent. However, bonding using an intermediate layer produces large thermo-mechanical stresses or presents microvoids between the bonded surfaces, which may degrade the bonding performance of the microfluidic device [47]. Nonetheless, high temperature bonding processes such as anodic bonding and fusion bonding are not applicable for polymeric materials. `Oxygen plasma` is widely used to permanently bond PDMS-PDMS, PDMS-Si or PDMS-glass layers and the bond is sufficiently strong to impede the two layers to be peeled. `Thermal compression` at T_{glass} and `adhesives` (e.g., resins, UV-curable materials, etc.) are used for polymer-to-polymer bonding. Although simple and low demanding, they may lead to localized deformation of the device and induced inhomogeneities in the interfacial layer thickness. Uniformity of the interfacial layer is achieved by other bonding techniques such as localized welding by ultrasounds [41], microwaves [20] or infrared laser [15] but bonding complexity is increased.

2.6 Summary

The demand for small, portable, and simple to use diagnostic devices to obtain fast or even instant diagnostic results is key driver for the development of a wide variety of microfluidic devices. Materials such as silicon, glass, polymers or even paper-based materials are being pursued. The current state-of-the-art of microfabrication techniques are significantly mature and have proven effective in detection applications in controlled environments. Nevertheless, there are a few challenges to overcome before these may become commercially established e.g., pre-treatment steps, concentration steps, reduction/elimination of dead volumes, elimination of external bulky and expensive detection systems, etc.

The present lack of a mass market for the relatively recent diagnostic devices using microfluidic technology makes it expensive to manufacture and therefore less competitive. This becomes more evident when these are compared with solutions already established in the market, with proven results and familiar to the handling personnel, which are the result of decades of continuous improvement and have already gone through all the regulatory approval procedures in different countries.

Apart from all the challenges, currently one of the main areas of application where microfluidic devices have been tested with observable success is in biomedical diagnostics. As a result, and as microfluidic devices become more common in the market, it is expectable to witness new areas of research and development of such devices for commercial use to become common.

In this chapter, the following 10 key remarks/tips have been provided:

#1 `Lithography` steps: (1) designing the pattern, (2) making the mask, (3) coating the wafer, (4) exposing the photoresist, (5) developing the photoresist.

#2 `Photoresist` is a light-sensitive polymeric solution. Negative PR (the regions exposed to light become insoluble in the developer) originate m.f.s. up to 2 µm and spin-coated spreading thickness from 1 µm to 3 µm (the photoresist film thickness is determined by the spinning speed), while positive PR (the regions exposed to light become soluble in the developer) has m.f.s. of $\sim$1 µm for spreading thicknesses between 1.5 to 7 µm.

#3 `Isotropic etch`: surface material is removed uniformly in all directions of the chemical structure. `Anisotropic etch`: the removal of surface material is dependent on the crystalline structure orientation of the surface.

#4 The etch rate in `wet etching` can be enhanced by tuning agitation and temperature.

#5 The `DRIE process` provides deep features of hundreds to thousands of micrometers into the substrate with nearly vertical walls.

#6 `Additive microfabrication` steps: (1) synthesize the vapor coating from the source (target) material, (2) transport the vapor to the substrate, (3) condense the vapor coating onto the surface of the substrate.

#7 The `PDMS` is composed of a two-part heat-curable mixture. The pre-polymer is cross-linked with the curing agent usually in a 10:1 ratio in weight, but varying this ratio one can obtain different mechanical and chemical properties of the resulting mixture.

#8 `Soft lithography` is the technique used to rapidly and easily fabricate and replicate a wide range of elastomeric devices, i.e., mechanically soft materials (e.g., polymers, gels or organic monolayers) without costly capital equipment.

#9 Heating and cooling ramps define the mechanical properties of the `molded part`.

#10 In general, the `bonding techniques` require flat, smooth and clean substrates to prevent voids between the bonded parts and achieve flawless bonding, even though they increase the costs of the overall fabrication and the material.

Acknowledgements The authors would like to thank Jorge M. Pereira for his contribution in the graphical editing of the document. This work has received funding from European Structural and Investment Funds through COMPETE Programme and from National Funds through FCT—*Fundação para a Ciência e a Tecnologia* under the Programme grant SAICTPAC/0019/2015. INESC-MN also acknowledges FCT funding through the Instituto de Nanociência e Nanotecnologia (IN) Associated Laboratory (UID/NAN/50024/2013).

References

1. Bashir, S., Bashir, M., Casadevall, I., Solvas, X., Rees, J. M., & Zimmerman, W.B. (2015). Hydrophilic surface modifcation of PDMS microchannel for O/W and W/O/W emulsions. *Micromachines, 6*, 1445–1458. doi:10.3390/mi6101429.
2. Bustillo, J. M., Howe, R. T., & Muller, R. S. (1998). Surface micromachining for microelectromechanical systems. *Proceedings of the IEEE, 86*(8), 1552–1574. doi:10.1109/5.704260.
3. Cardoso, S., Leitao, D. C., Dias, T. M., Valadeiro, J., Silva, M. D., Chicharo, A., et al. (2017). Challenges and trends in magnetic sensor integration with microfluidics for biomedical applications. *Journal of Physics D: Applied Physics*. doi:10.1088/1361-6463/aa66ec
4. Carrilho, E., Martinez, A. W., & Whitesides, G. M. (2009). Understanding wax printing: A simple micropatterning process for paper-based microfluidics. *Analytical Chemistry, 81*, 7091–7095.
5. Cheng, C., Ke, K.-C., & Yang, S.-Y. (2017). Application of graphene polymer composite heaters in gas-assisted micro hot embossing. *RSC Advances, 7*, 6336–6344. doi:10.1039/C6RA27618K.
6. Duan, X., Zhao, Y., Perl, A., Berenschot, E., Reinhoudt, D. N., & Huskens, J. (2010). Nanopatterning by an integrated process combining capillary force lithography and microcontact printing. *Advanced Functional Materials, 20*, 663–668.
7. Fernandes, J. T. S., Chutna, O., Chu, V., Conde, J. P., & Outeiro, T. F. (2016). A novel microfluidic cell co-culture platform for the study of the molecular mechanisms of parkinson's disease and other synucleinopathies. *Frontiers in Neurosciences, 10*, 511. doi:10.3389/fnins.2016.00511.
8. Fu, E., & Downs, C. (2017). Progress in the development and integration of fluid flow control tools in paper microfluidics. *Lab on a Chip*. doi:10.1039/c6lc01451h.
9. Goral, V. N., Hsieh, Y.-C., Petzold, O. N., Faris, R. A., Yuen, P. K. (2011). Hot embossing of plastic microfluidic devices using poly(dimethylsiloxane) molds. *Journal of Micromechanics and Microengineering, 21*(1):017002 (8 pages). doi:10.1088/0960-1317/21/1/017002
10. Guckenberger, D. J., de Groot, T. E., Wan, A. M. D., Beebe, D. J., & Young, E. W. K. (2015). Micromilling: A method for ultra-rapid prototyping of plastic microfluidic devices. *Lab on a Chip, 15*(11), 2364–2378. doi:10.1039/c5lc00234f.
11. Halldorsson, S., Lucumi, E., Gmez-Sjberg, R., & Fleming, R. M. T. (2015). Advantages and challenges of microfluidic cell culture in polydimethylsiloxane devices. *Biosensors and Bioelectronics, 63*, 218–231.
12. Hwang, H., Seo, D., Park, J., & Park, C. (2014). Investigation of the power transistor size related to the efficiency of switching-mode RF CMOS power amplifier. *Microwave and Optical Technology Letters, 56*(1), 110–17. doi:10.1002/mop.
13. Kaufmann, T., & Ravoo, B. J. (2010). Stamps, inks and substrates: polymers in microcontact printing. *Polymer Chemistry, 1*, 371–387.
14. Kecskemeti, A., Bako, J., Csarnovics, I., Csosz, E., & Gaspar, A. (2017). Development of an enzymatic reactor applying spontaneously adsorbed trypsin on the surface of a PDMS microfluidic device. *Analytical and Bioanalytical Chemistry*. doi:10.1007/s00216-017-0295-9.
15. Kim, J., & Xu, X. (2003). Excimer laser fabrication of polymer microfluidic devices. *Journal of Laser Applications, 15*, 255.
16. Konstantinou, D., Shirazi, A., Sadri, A., & Young, E. W. K. (2016). Combined hot embossing and milling for medium volume production of thermoplastic microfluidic devices. *Sensors and Actuators B: Chemical, 234*, 209–221.
17. Kuo, C.-J., Peles, Y. (2009). Flow boiling of coolant (HFE-7000) inside structured and plain wall microchannels. *Journal of Heat Transfer, 131*:121011, 9 pages. doi:10.1115/1.3220674.
18. Kusaka, Y., Miyashita, K., & Ushijima, H. (2014). Extending microcontact printing for patterning of thick polymer layers: semi-drying of inks and contact mechanisms. *Journal of Micromechanics and Microengineering, 24*(12).

19. Lee, T. M. H., Lee, D. H. Y., Liaw, C. Y. N., Lao, A. I. K., & Hsing, I.-M. (2000). Detailed characterization of anodic bonding process between glass and thin-film coated silicon substrates. *Sensors and Actuators*, *86*, 103–107. doi:10.1016/S0924-4247(00)00418-0.
20. Lei, K. F., Ahsan, S., Budraa, N., Li, W. J., & Mai, J. D. (2004). Microwave bonding of polymer-based substrates for potential encapsulated micro/nanofluidic device fabrication. *Sensors and Actuators A*, *114*, 340–346.
21. Li, C., Boban, M., & Tu, A. (2017). Open-channel, water-in-oil emulsification in paper-based microfluidic devices. *Lab on a Chip*, *17*, 1436–1441. doi:10.1039/C7LC00114B.
22. Lin, T.-Y., Do, T., Kwon, P., & Lillehoj, P. B. (2017). 3D printed metal molds for hot embossing plastic microfluidic devices. *Lab on a Chip*, *17*, 241–247. doi:10.1039/C6LC01430E.
23. Linder, P., Dragoi, V., Farrens, S., Glinsner, T., & Hangweier, P. (2004). Advanced techniques for 3D devices in wafer - bonding processes. *Solid State Technology*, *47*(6), 55–58.
24. Liu, K., & Fan, Z. H. (2011). Thermoplastic microfluidic devices and their applications in protein and DNA analysis. *Analyst*, *136*(7), 1288–1297.
25. Luttge, R. (2011). *Microfabrication for industrial applications*. USA: Elsevier Inc. ISBN 987-0-8155-1582-1.
26. Madou, M. J. (2012). *Manufacturing Techniques for Microfabrication and Nanotechnology* (3rd ed.). Boca Raton, USA: CRC Press. ISBN 978-0849331800.
27. Menz, W., Mohr, J., Paul, O. (2008). *Microsystem technology*. John Wiley and Sons, Inc., ISBN: 978-3-527-61301-4.
28. Mou, L., & Jiang, X. (2017). Materials for Microfluidic Immunoassays: A Review. *Advanced Healthcare Materials*, 1601403. doi:10.1002/adhm.20160140.
29. Nguyen, N.-T., Wereley, S. T. (2002) *Fundamentals and applications of microfluidics*. Artech House Publishers, ISBN 10: 1580533434, ISBN 13: 9781580533430.
30. Peng, L., Deng, Y., Yi, P., Lai, X. (2014). Micro hot embossing of thermoplastic polymers: a review. *Journal of Micromechanics and Microengineering*, *24*:013001 (23pp). doi:10.1088/0960-1317/24/1/013001.
31. Posthuma-Trumpie, G. A., Korf, J., & van Amerongen, A. (2009). Lateral flow (immuno)assay: its strengths, weaknesses, opportunities and threats. A literature survey. *Analytical Bioanalytical Chemistry*, *393*, 569–582.
32. Rai-Choudhury, P. (1997). *Handbook of Microlithography, Micromachining, and Microfabrication* (Vol. 1) Microlithography. Spie Press Book, ISBN: 9780819497864
33. Ramm, P., Lu, J. J.-Q., & Taklo, M. M. V. (2012). *Handbook of wafer bonding*. KGaA, Germany: Wiley-VCH Verlag GmbH and Co. ISBN 978-3-527-32646-4.
34. Sathish, S., Ricoult, S., Toda-Peters, K., & Shen, A. Q. (2017). Microcontact printing with aminosilanes: Creating biomolecule micro- and nanoarrays for multiplexed microfluidic bioassays. *Analyst*. doi:10.1039/C7AN00273D.
35. Schwartz, B., & Robbins, H. (1976). Chemical etching of silicon, IV. Etching technology. *Journal of the Electrochemical Society*, *123*(12), 1903–1909. doi:10.1149/1.2132721.
36. Shamsi, A., Amiri, A., Heydari, P., Hajghasem, H., & Mohtashamifar, M. (2014). Esfandiari M (2014) Low cost method for hot embossing of microstructures on PMMA by SU-8 masters. *Microsystem Technologies*, *20*, 1925. doi:10.1007/s00542-013-2000-z.
37. Shangguan, J.-W., Liu, Y., Pan, J.-B., Xu, B.-Y., Xu, J.-J., & Chen, H.-Y. (2017). Microfluidic PDMS on paper (POP) devices. *Lab on a Chip*, *17*, 120–127. doi:10.1039/C6LC01250G.
38. Silverio, V., Cardoso, S., Gaspar, J., Freitas, P. P., & Moreira, A. L. N. (2015). Design, fabrication and test of an integrated multi-microchannel heatsink for electronics cooling. *Sensors and Actuators A: Physical*, *235*, 14–27.
39. Soares, R. R. G., Novo, P., Azevedo, A. M., Fernandes, P., Aires-Barros, M. R., Chu, V., et al. (2014). On-chip sample preparation and analyte quantification using a microfluidic aqueous two-phase extraction coupled with an immunoassay. *Lab on a Chip*, *14*, 4284–4294. doi:10.1039/C4LC00695J.
40. Stamps, R. L., Breitkreutz, S., kerman, J., Chumak, A. V., Otani, Y., Bauer, G. E. W. et al. (2014). The 2014 magnetism roadmap. *Journal of Physics D: Applied Physics*, *47*:333001 (28pp).

41. Truckenmuller, R., Ahrens, R., Cheng, Y., Fischer, G., & Saile, V. (2006). An ultrasonic welding based process for building up a new class of inert fluidic microsensors and actuators from polymers. *Sensors and Actuators A, 132*(1), 385–392.
42. Tsao, C.-W. (2016). Polymer microfuidics: simple, Low-Cost Fabrication Process Bridging Academic Lab Research to commercialized production. *Micromachines, 7*, 225. doi:10.3390/mi7120225.
43. Voicu, D., Lestari, G., Wang, Y., DeBono, M., Seo, M., Cho, S., et al. (2017). Thermoplastic microfluidic devices for targeted chemical and biological applications. *RSC Advances, 7*, 2884–2889. doi:10.1039/C6RA27592C.
44. Yang, Y., Noviana, E., Nguyen, M. P., Geiss, B. J., Dandy, D. S., & Henry, C. S. (2017). Paper-based microfluidic devices: emerging themes and applications. *Analytical Chemistry, 89*(1), 71–91.
45. Young, E. W. K., Berthier, E., & Beebe, D. J. (2012). Assessment of enhanced autofluorescence and impact on cell microscopy for microfabricated thermoplastic devices. *Analytical Chemistry, 85*, 44–49. doi:10.1021/ac3034773.
46. Yu, L., & Shi, Z. Z. (2015). Microfluidic paper-based analytical devices fabricated by low-cost photolithography and embossing of Parafilm, *Lab on a Chip, 15*, 1642–1645. doi:10.1039/C5LC00044K.
47. Zhang, H., Zhang, Q., Chong, S.-C., Pinjala, D., Liu, X., Chan, P.-K. (2006) Development and characterization of large silicon microchannel heat sink packages for thermal management of high power microelectronics modules. In *Proceedings of the 56th Electronic Components and Technology Conference* (pp. 1018-1022). doi:10.1109/ECTC.2006.1645778.
48. Zhang, H., & Chiao, M. (2015). Anti-fouling coatings of poly(dimethylsiloxane) devices for biological and biomedical applications. *Journal of Medical Biological Engineering, 35*(2), 143–155.

Chapter 3
Fluid-Flow Characterization in Microfluidics

Laura Campo-Deaño

Abstract The number of scientific studies related with microfluidics techniques have been increasing in the last decade due to the great advantages they provide in the characterization of complex fluid flows. Obtaining precise and accurate results, especially in the experimental field, is of great importance. However, this is not always an easy task. In this chapter we analyze fundamental aspects related to the experimental techniques in microfluidics, providing also some useful tips to assist anyone who decides to start in this exciting "micro world". A brief introduction about the potentialities of microfluidics will be presented. Additionally, a small review of the most common experimental techniques and their advantages and disadvantages are also assessed. Finally, an overview of the most common errors/difficulties of these experimental techniques in the lab is introduced.

3.1 Introduction

As it was stated along this book, microfluidics is a powerful technology based on the manipulation of tiny volumes of fluid. This technique constitutes a new alternative to study and solve some fluid flow complex problems in which the conventional macro techniques are not able to reach. Particularly in the area of biomedicine, to provide new assays system for blood characterization, blood separation, drug delivery, but also as an useful tool when the available material to test, i.e. the amount of fluid, is very limited or costly.

As a first step to understand why the complex fluid flow at micro scale is so different from the complex fluid flow at macro scale, we will examine the most important dimensionless numbers that play a very important role. One of them is the Reynolds number, which is the relation between the inertial and viscous forces and can be expressed as $Re = (\rho VL)/\eta$, where η and ρ are the viscosity and the density of the fluid, respectively, V the average velocity and L the characteristic length of

L. Campo-Deaño (✉)
Department of Mechanical Engineering, Faculty of Engineering of the University of Porto,
Rua Dr. Roberto Frias s/n, 4200-465 Porto, Portugal
e-mail: campo@fe.up.pt

© Springer International Publishing AG 2018
F.J. Galindo-Rosales (ed.), *Complex Fluid-Flows in Microfluidics*,
DOI 10.1007/978-3-319-59593-1_3

the micro channel. Considering water as working fluid, velocities of the order of 1 μm/s or cm/s and typical length scales of 100 or 1 μm, the Reynolds number will vary between 10^{-6} and 1, and it is possible to affirm that viscous forces typically dominate over inertial forces. Therefore, the flow in micro devices can be considered laminar even when the working fluid presents very low viscosity [20, 27].

When dealing with complex fluids, another important parameter is the Weissemberg number, which account for the relation of the relaxation time of the fluid (λ) and the flow deformation time, $Wi = \lambda(V/L)$, being λ the relaxation time of the fluid. On the other hand, another similar dimensionless parameter is the Deborah number, defined as the relation between the relaxation time of the fluid and the time of observation of the flow (t), $De = \lambda/t$, that helps (as in the case of the Wi number) to characterize flows presenting some degree of elasticity apart, obviously, from viscosity. In microfluidics, the combination of the high deformation rates, small time scale and small characteristic length leads to attain high De or Wi numbers meanwhile the Re is relatively low. This fact favors the observation of strong viscoelastic effects that, in the case of very low viscoelastic fluids, would be hided by inertial effects at the macro scale [20].

Taking these considerations into account, it is easy to understand how the complex flows in micro scale are too different from the complex flows at macro scale.

In microfluidics, the micro channels can be either designed for mimicking an specific geometry to reproduce *in vitro* some natural or biological phenomena [5, 15, 17, 24], to study the flow dynamics around different bodies [4, 14] or specially designed to analyze the viscoelastic properties of complex fluids, i.e. rheometer-on-a-chip (Fig. 3.1).

In this latter case, several progress have been performed in the last years [2, 11, 19, 22, 33]. Pipe and McKinley [23] wrote an interesting review about the use of microfluidics techniques to measure rheological properties in shear and extensional flows. In a more recent review Galindo-Rosales et al. [10] present a nice overview of several micro channel designed to characterize the extensional properties of fluids. More information about microrheometry can be found in Chap. 1.

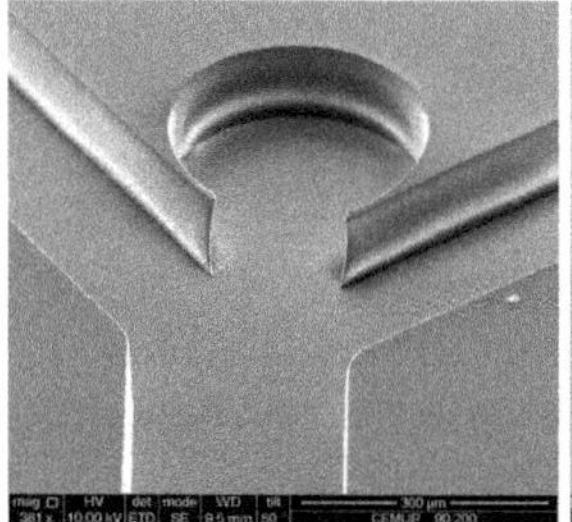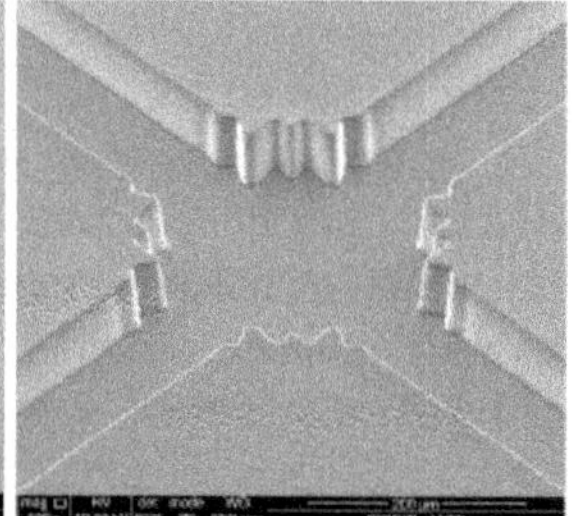

Fig. 3.1 Scanning Electron Microscopy (SEM) images for different geometry micro channels. *Left* simplified 2D micro channels mimicking an aneurysm; *centre* 3D microbot prototypes; *right* optimized cross-slot geometry for generating pure extensional flows

In order to characterize the main features of the complex flows in micro scale it is very common to use different experimental techniques. These techniques are able to provide the velocity profiles along the micro channels, the streamlines, pressure drop differences and even the stress distribution. A more detailed review of these experimental techniques will be presented in the following section.

The contents in this chapter book are as follows: after the Introduction, Sect. 3.2 will deal with the experimental techniques commonly used in microfluidics for the characterization of complex fluid-flows; in Sect. 3.3, some tips on how to solve possible experimental problems will be presented, and finally, some remarks are explained in Sect. 3.4.

3.2 Experimental Techniques

The working fluids used in microfluidics are normally transparent, making impossible to analyze the flow pattern as the fluid motion is invisible for us. In order to overcome this limitation, different methods can be considered, such as visualizing the surface flow (i.e. adding colored oil to the surface channel), analyzing the changes in the refractive index of the flow or by adding tracer particles to the flow. The latter one is the most common technique used in microfluidics and hereafter we will refer to this methodology. The nature and size of the tracer particles will depend on the characteristic of the fluid, the morphology of the micro channels and also on the intrinsic characteristic of the experimental technique. In general, the tracer particles must to present the following characteristics: their density should be close to the density of the working fluid; they have to be non-volatile, non-toxic or corrosive and they should not undergo any chemical reaction; and they should present an optimum size [7].

A proper design of the experimental set up and an accurate assembly of all its components is crucial for a precise analysis of the flow behavior in micro channels. The equipment of a complete set up varies according the experimental technique, however as a general rule a microfluidic setup for the study of the flow behavior of complex fluid flows must be composed of the following items:

- Inverted microscope equipped with a CCD camera, a light source and a filter cube.
- Syringe pump to inject the fluid and control the flow rate in the micro channel.
- Syringes with different volumes.
- Tubes for connecting the syringes with the channels.
- Connectors, valves and tips.
- Micro channel geometries.
- Working fluids.
- Computer.

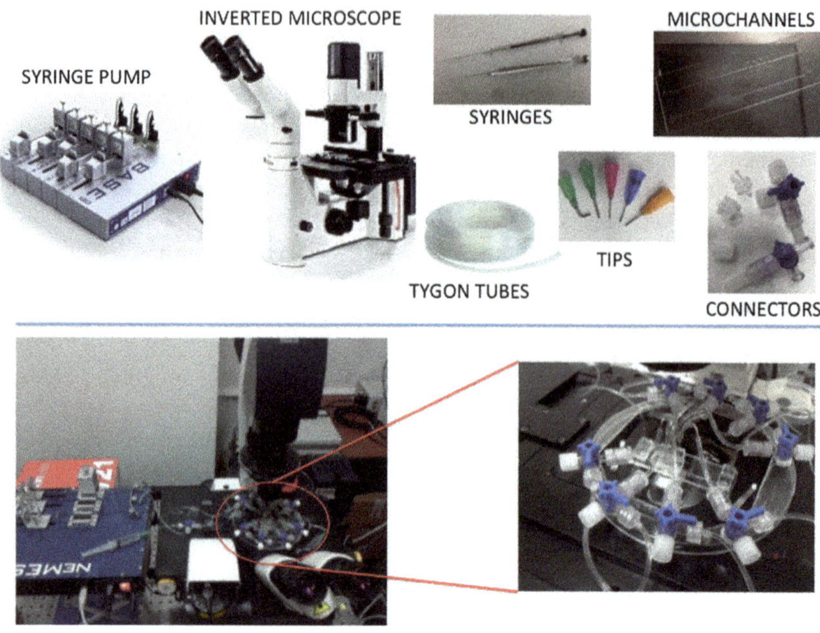

Fig. 3.2 General setup for a microfluidic experiment. The syringe pump image is courtesy from CETONI, GmbH. This modular pump offers an extremely precise and stable flow with virtually no pulsation. Image of the inverted microscope is courtesy from Leica Microsystems

> A proper design of the microfluidic set up is crucial for obtaining reliable and accurate measurements of the main flow characteristics.

Example of a general microfluidic set up is shown in Fig. 3.2. The main components as the syringe pump, the microscope, syringes, supply tubes and micro channels are essential independently of the experimental technique.

More detail about the properties of each component of the setup will be covered in the following subsections for each particular technique. In Fig. 3.3 the most common flow characterization techniques are summarized according to their applicability. It is necessary to point out that these techniques are non-invasive and therefore its applicability does not interfere in the rheological properties of the fluid either in the fluid flow behavior.

> Knowing the potentialities and applicability of the available experimental techniques is crucial for the selection of the most appropriate method for fluid flow characterization.

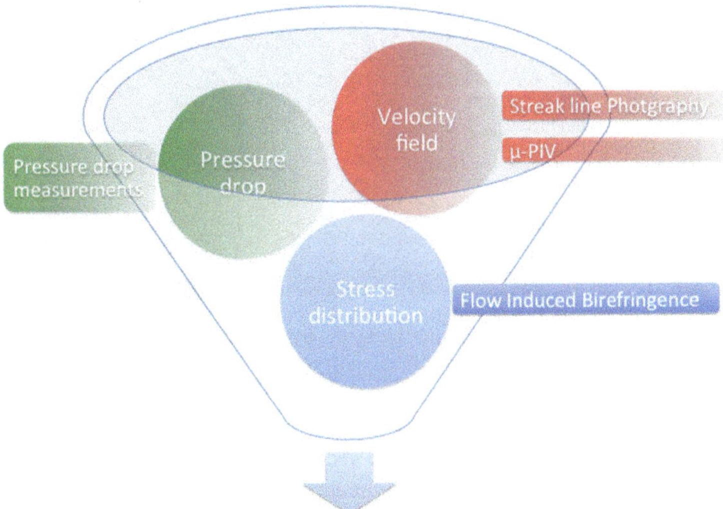

Fig. 3.3 Overview of the main experimental techniques for fluid flow characterization in microfluidics

3.2.1 Streak Line Photography

Streak line photography is a flow visualization technique that can be applied either at macro or at micro scale. It consists on recording particles displacements over a period of time, allowing a qualitative analysis of the flow pattern. The fluids should be seeded with fluorescent tracer particles, it is also recommendable to add an small amount of surfactant, i.e. Sodium Dodecyl Sulfate (SDS), around 0.05 wt%, to diminish adhesion of the tracer particles to the micro channel walls.

Fig. 3.4 Mercury lamp illumination during a streak line photography experiment

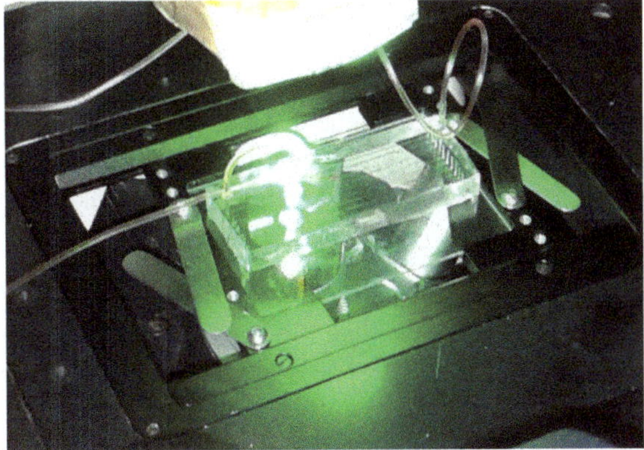

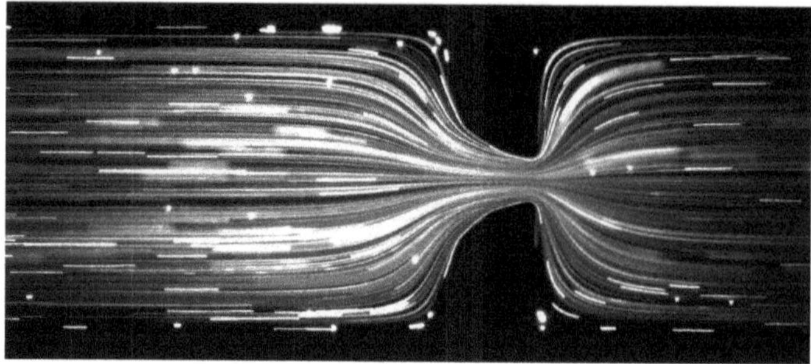

Fig. 3.5 Flow visualization using streak line photography technique of a viscoelastic fluid flowing through a contraction-expansion geometry

The micro channels which contain the seeded fluid is continuously illuminated using a mercury lamp (Fig. 3.4) and the light emitted by the tracers is imaged through the objective lens onto the CCD array of the camera using long exposure times to capture the pathlines of the particles. The exposure times should be varied accordingly with the flow rate and the dimensions of the channels. Figure 3.5 shows a viscoelastic fluid flow through a contraction-expansion geometry using streak line photography.

3.2.2 Micro—Particle Image Velocimetry (μ-PIV)

Particle image velocimetry is an imaging method normally used for quantifying the velocity field which results from the measure of the local velocity based on the average displacement of fluorescent particles in an interrogation area over a known time. The motion of these particles within each interrogation region is determined using cross correlation algorithms [30].

For its use in microfluidics, it is normally to perform a μ-PIV technique which presents some fundamental differences from the macro-PIV technique. In conventional PIV a light sheet is used to illuminate the tracer particles, while in μ-PIV a volume illumination is applied and interactions between particles and particles-fluid can originate Brownian motion due to their small size. The spatial resolution of this experimental technique depends on the tracer particle size, tracer particle density and image quality, among others [25]. In order to perform accurate micro-PIV measurements some considerations must be taking into account according to [18], mainly:

- *In-plane spatial resolution*: the tracer particles used in this technique should be imaged with enough high magnification, about three-four pixels in diameter. The diffraction limited spot size of a point source of light d_s is as follows:

$$d_s = 2.44(M + 1)\frac{\lambda}{2NA} \tag{3.1}$$

where M is the magnification of the objective λ is the wavelength of the light emitted or scattered by the tracer particles and NA the numerical aperture of the lens. The effective image diameter can be expressed as:

$$d_e = [d_s^2 + M^2 d_p^2]^{1/2} \tag{3.2}$$

being d_p the real particle diameter. When Md_p is smaller than d_s the diffraction effects dominate and d_s remains constant. If the image size is larger than d_s, then $d_e \approx Md_p$.

- *Out-of-plane spatial resolution*: the measurement plane is normally determined using the depth of field of the objective lens used. As the flow field is volume-illuminated in micro-PIV techniques, particles that are out-of-focus should be considered as they are contributing for the background noise. So, the thickness of the measurement plane, δz_m, (Fig. 3.6) is obtained taking into account three different effects: one due to geometrical optics, another one due to the finite size of the particle and finally, the effect of the diffraction (Eq. 3.3) [26]

$$\delta z_m = \frac{3n\lambda_0}{(NA)^2} + \frac{2.16 d_p}{tan\theta} + d_p \tag{3.3}$$

being n the refractive index, λ_0 is the wavelength of light, d_p the particle diameter and $tan\theta \approx sin\theta = NA/n$.

As it was stated above, due to the application of a volume illumination there are some unfocused particles that are contributing for the background noise. This noise is very difficult to remove as it normally presents the same wavelength as the focused particle. Then, the particle visibility, which is the ratio of the intensity of an in focus particle image to the background light produced by the out of focus particles, can be determined by the theory of Olsen and Adrian [21]. They found that particle visibility is increased by decreasing particle concentration or by decreasing the thickness of the test section. Also, the visibility would increase decreasing the particle diameter or increasing the NA of the lens, keeping constant the concentration of particles. They derived an equation for the volume fraction, V_{fr} of particles that produces a particular visibility (V):

$$V_{fr} = \frac{2d_p^3 M^2 \beta^2 (s_0 - L/2)(s_0 - L/2 + L)}{3V L s_0^2 [M^2 d_p^2 + 1.49(M+1)^2 \lambda^2 [(n/NA)^2 - 1]]} \tag{3.4}$$

where s_0 is the objective working distance and L the depth of the micro channel. $\beta^2 = 3.67$ defines the cut-off level of the edges of the particle. A visibility of about 1.5 is required for high quality velocity measurements.

- *Effects of Brownian motions*: the Brownian motion undergone by the tracer particles when they are very small can provoke errors in the velocity measurements and also can lead to a uncertainty in the location of the particles. As a consequence

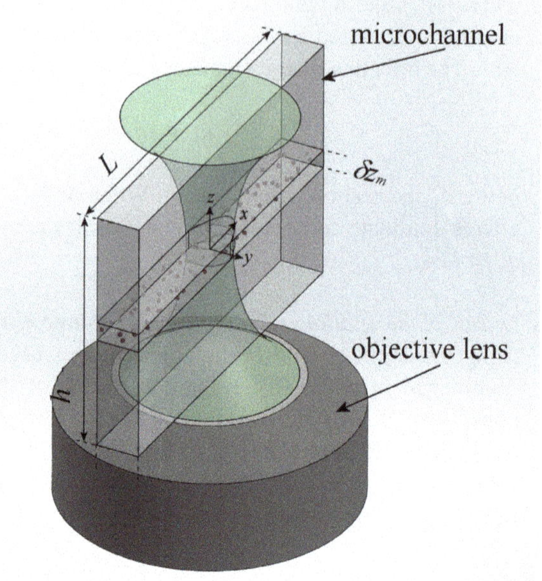

Fig. 3.6 Schematic representation of the volume illumination and the measurement plane, δz_m. Image courtesy of Simon J. Haward (OIST, Japan)

of imaging the Brownian particle displacements, the relative errors in x and y components of the velocity are:

$$\varepsilon_x = \frac{\sigma_x}{\Delta x} = \frac{1}{u}\sqrt{\frac{2D}{\Delta t}} \tag{3.5}$$

$$\varepsilon_y = \frac{\sigma_y}{\Delta y} = \frac{1}{u}\sqrt{\frac{2D}{\Delta t}} \tag{3.6}$$

where σ is the displacement of the particles in each direction, Δx and Δy are the total particle displacement, u and v are the velocity components in x and y directions, Δt is the time interval between images and D is the diffusion coefficient:

$$D = \frac{k_B T}{3\pi \mu d_p} \tag{3.7}$$

being k_B the Boltzmann's constant ($k_B = 1.3807 \times 10^{-23}$ m^2 kg/ s^2 K), T the absolute temperature and μ the dynamic viscosity of the fluid [28].

Therefore, it is possible to affirm that the effect of the Brownian motion of the particles is very low for faster flows.

3.2.3 Pressure Drop Measurements

This experimental technique provides valuable information in microfluidics. Different pressure sensors covering various differential pressure ranges can be employed according to the pressure drop one wanted to measure, which depends on the channel characteristics and the flow rates under study. The most common pressure sensors used in this area are capacitive or piezoresistive. In the last years, piezoresistive sensors systems with high preciseness and sensitivity for pressure measurements have been developed, and they are able to measure absolute and relative pressure (Fig. 3.7) [6, 8].

The differential pressure sensors, normally used in microfluidics, should be calibrated utilizing two different methods: a static column of water for sensors covering small pressure differences and a compressed air line with a manometer for sensors covering higher differential pressures. For that end, two pressure ports must be located upstream and downstream of the area of study. The pressure sensors are powered by a power supply and connected to a computer using a data acquisition card to register the data (voltage). Figure 3.8 shows a typical calibration curve for a differential pressure sensor, it is clear a linear dependence of the voltage with the imposed pressure difference in a water column ($\Delta P = \rho g \Delta h$).

The pressure sensors typically show a transient response until steady-state is reached. For Newtonian fluids, the steady-state is reached more rapidly than for viscoelastic fluids (Fig. 3.9). After steady-state is attained, the signal reaches a plateau with superimposed low amplitude oscillations due mostly to electronic noise, and the average pressure drop is calculated by taking the arithmetic mean of the signal with time [2].

Fig. 3.7 Image of a differential pressure sensor used in microfluidics

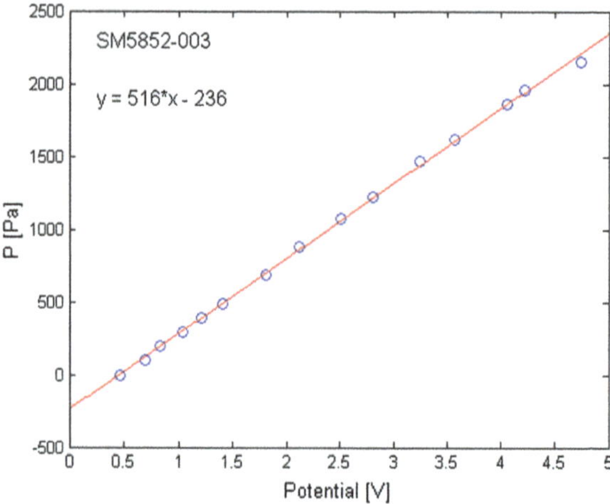

Fig. 3.8 Typical calibration *curve* for a differential pressure sensor

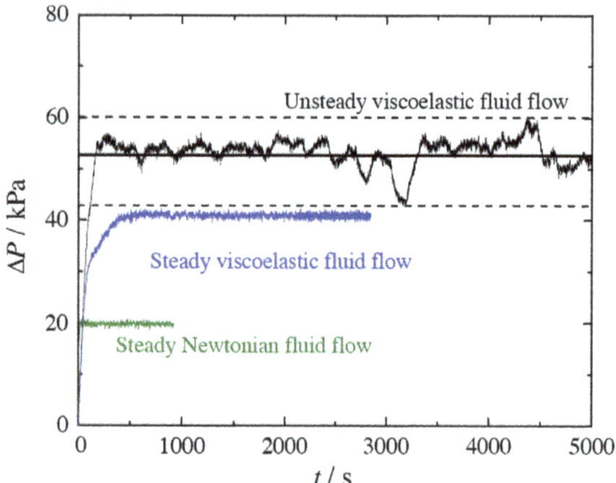

Fig. 3.9 Pressure drop measurements as a function of time for Newtonian and viscoelastic fluids. Image courtesy of Patrícia C. Sousa (FEUP, Portugal)

3.2.4 Birefringence Techniques

Flow induced birefringence (FIB) is a non-invasive technique for measuring the stress distribution in polymer solutions. FIB is caused by small changes in the refraction of light passing through a fluid flow, and these changes correlate to elastic stresses in the flowing fluid [12]. The level of birefringence is higher as the polymer chains are elongated which implies higher stresses [32].

> The applicability of FIB techniques requires the use of birefringent materials introducing a restriction to the fluids being used.

For obtaining an optical birefringence pattern, the micro channels containing the working fluid are continuous illuminated using a monochromatic light. A helium-neon laser is normally used to allow the beam passed orthogonally through the micro channel. The laser is detected by a CCD camera recording at the area of interest. On each side of the micro channel, a polarizer and a analyser are located with axes at 45° to the flow axis, and at 90° to each other in order to diminish the signal taken by the camera. A $\lambda/4$ plate, which is a retardation plate causing a phase difference of $\pi/2$, is used to balance the residual birefringence in the micro channel and is placed within the polarizer and analyser. Around 20–25 ms images are normally captured at different flow rates. At the moment at which the flow is stopped, the background image is captured and removed from the images obtained under flow to show the birefringent signal. The intensity of the signal is calibrated by means of a $\lambda/30$ compensator in the optical line, when the compensator is rotated the intensity of the signal is obtained in order to get a range of retardation values. The relation between the retardation, R, and the birefringence, Δn, is as follows: $R = \Delta n\, l$, where l is the path-length through the material.

The normal stress and the shear stress can be obtained by means of the stress-optic law [16]. Equations 3.8 and 3.9 provides the relationship between the index of refraction tensor and the stress tensor:

$$\sigma_{12} = \frac{1}{2C}\Delta n sin(2\alpha), \tag{3.8}$$

$$\sigma_{11} - \sigma_{22} = \frac{1}{C}\Delta n cos(2\alpha), \tag{3.9}$$

being C the stress-optic coefficient which is independent of molecular weight, polydispersity, and polymer concentration [13, 29].

Birefringence measurements provide the stress tensor directly only when the flow is considered to be 2D. Otherwise, for a 3D flow, the birefringence and the orientation angle vary along the light path leading to a cumulative measurement of the optical properties [32].

3.3 Some Typical Errors/Difficulties

It is normal that before and during the experimentation different kind of difficulties appear. In this section, some of these problems are summarized as well as some tips in order to help to overcome these complications.

3.3.1 Optical Distortions

Sometimes, the necessity of reproducing exactly some typical geometries, as in the case of the human circulatory system, instead of using rectangular cross-section micro channels it is common to use circular cross-section micro channels. In this latter case, an special attention should be considered for matching the refractive index of the working fluid and of the channel in order to diminish optical distortions, hidden black regions and reflections that would affect the measurements of the experimental techniques. An schematic representation of the differences between a geometry with solid-liquid interfaces with curvature and without curvature is presented in Fig. 3.10, where a clear deviation of the rays is observed in the case of the circular geometry.

The real effect can be observed in Fig. 3.11, a circular cross-section milichannel made of PDMS is filled with different fluids having different refractive indices. For a refractive index of 1.33, much lower than that of PDMS, a wide black region close to the walls appears leading to a confusing detection of particle positions. If we increase the refractive index, close to that of the PDMS (1.41) [3], the distortion diminishes, allowing the correct characterization of the fluid-flow.

> In the case of micro channels with a circular cross-section geometry the refractive index of the working fluid should match the refractive index of the micro channel material in order to avoid optical distortions.

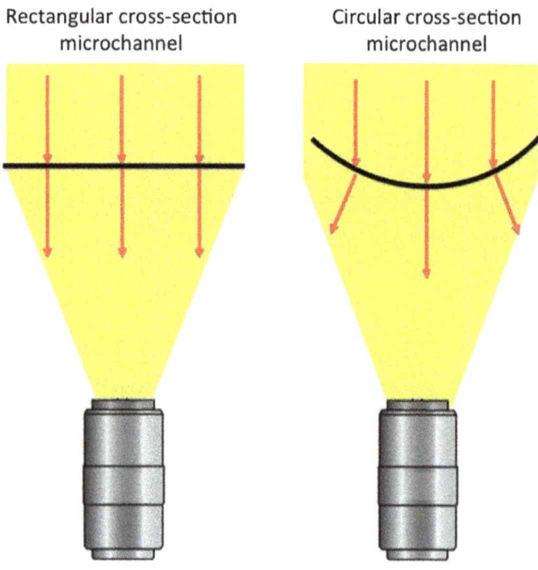

Fig. 3.10 Schematic representation of the refraction of light in a rectangular cross-section and in a circular cross-section micro channel

Fig. 3.11 Microscope images in the *middle plane* of a PDMS milli-channel with circular cross-section filled with solutions with different refractive index: 1.33 (*left*), 1.39 (*middle*), and 1.41 (*right*) containing 15.47 μm spherical particles (refractive index of PDMS was 1.41). Reproduced from [3], with the permission of AIP Publishing

3.3.2 Entrance Length and Exit Effects

When the working fluid is Newtonian the calculation of the entrance length (L_e) dimension is trivial, either in laminar or turbulent flows, although this latter case is not usual in microfluidics. However, when the fluid flow is non-Newtonian the characterization of the point at which the flow becomes totally developed is not an easy task. Elastic effects play a very important role and the determination of their influence in the entrance length is not well stablish until data. Only evaluating the velocity profile at different points along the micro channel will help to define this important length.

Li and Haward [31] investigated the flow of a Newtonian fluid and a viscoelastic fluid through a micro fabricated capillary entrance device. The concluded that the effect of the elasticity significantly increases the entrance length and it depends on the Wi number.

> The determination of the entrance length when the working fluid is non-Newtonian is not trivial. This distance could be only determined by measuring the velocity profiles along the micro channel.

The end effects are also very important and, as in the case of the entrance length, the distance between the exit and the area of interest must be long enough to avoid instabilities. The experimental determination of the velocity profile along the channel is the most accurate way to delimit this length.

The entrance and exit effects are particular important when the pressure drop need to be determined. There is an excess of pressure drop that can be corrected using the well-known Bagley correction [1].

3.3.3 Presence of Bubbles in Tubes and Channels

When the flow is injected into the micro channel through the syringe and the supply tubes, it is very common to notice the rise of some air bubbles. The presence of these bubbles affects the accuracy and reliability of the measurements. In order to solve this problem several options can be considered:

- In a first step the micro channels should be washed with ethanol and deionized water, after that the working fluid should be introduced [19].
- Increase the flow rate and keep the flow during long time. In the case of micro channels made of PDMS, as this material is porous it would help in the elimination of the air bubbles.
- Fill in completely the tubes with the fluid before connecting the syringe to the tube, the same with the syringe. In this way, no bubbles would enter in the instant of the connection.

> The rising of air bubbles affects the flow and avoids the obtention of accurate and reliable measurements. It is important to verify that the micro channel is free of air bubbles.

3.3.4 Sedimentation of Particles

In the case of suspension in which the particles of the solution act as tracer particles, i.e. the study of flow behavior of cells in a viscoelastic medium, it is important to avoid the sedimentation of the particles. The sedimentation could appear either in the supply tubes or in the microchanel. In micro channels, taking into account the flow transient times this effect could be negligible being only important at very low flow rates. However, the sedimentation in the supply tubes and in the syringe could be more significant due to the long time the working fluid is inside the syringe during the experiment. Matching the density of the particle with the density of the solvent fluid is essential to reduce this problem.

The sedimentation is enhanced if the particles tend to form agglomerates. The addition of a surfactant, as Sodium Dodecyl Sulfate (SDS), in small amount around 0,05% helps to avoid the agglomeration. Dextran 40 is very commonly used in *in vitro* blood flow experimental studies as a substitute of blood plasma to diminish the sedimentation and clogging of the cells.

In the case of using fluorescence tracer particles, the proper selection of the size and concentration of these particles is crucial for accurate measurements. This aspect was already addressed in Sect. 3.2.2. However, other important phenomenon related to the adhesion of these particles in the micro channel walls must be also considered here. Figure 3.12 shows an example of an excess of light due to the concentration of

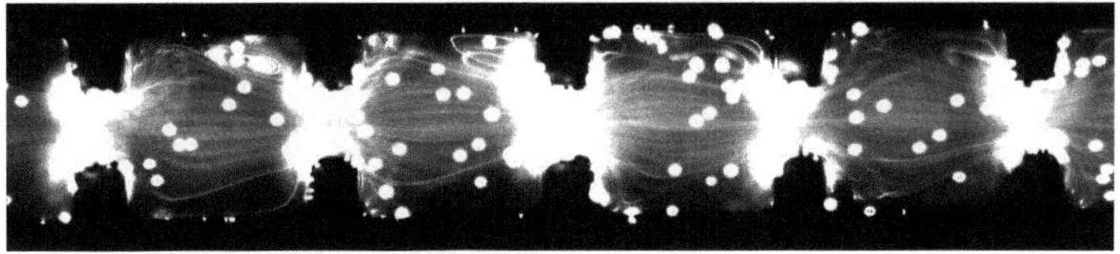

Fig. 3.12 Effect of the excited-state saturation of the tracer particles as a consequence of the stuck of the tracers in the channel walls

tracer particles in the walls of the micro channel. This excess of light is a consequence of excited-state saturation of the particles as they remain fixed to the walls during time. The addition of a surfactant will decrease considerably this undesirable effect. However, one should be careful with the addition of surfactant, as we may modify the rheological behaviour of our sample.

> Sedimentation, aggregation and stuck of tracer particles in the channel or in the supply tubes should be avoided by adding specific solutions and by matching the density between particles and the working fluid.

3.3.5   Focusing the Mid Plane

One of the fundamental problems at the time to perform a microfluidic characterization technique, is the location of the mid plane. When the lens used is of low magnification, the problem is not important as the field of view of the lens is sometimes higher than the depth of the microchanel. However, when the magnification lens is high, it is crucial to find the focussing plane. In order to find the proper location of the focussing plane, the bottom part of the channel must be focused using the adjustment knob of the microscope (Fig. 3.13) trying to get the sharpest possible image, after that the lens is moved away in order to focus the top part of the micro channel and we proceed in the same way. The mid plane corresponds with the medium distance between these two focused walls.

> Before performing any flow characterization technique, the focussing of the mid plane must be guaranteed.

Fig. 3.13 Knob of the microscope used for the adjustment of the focussed image

3.3.6 Pulses in Syringe Pumps

The pulses in the flow associated with the syringe pump when the flow rate is imposed are critical to reliable operation of microfluidic devices and could lead to undesired oscillations in the flow rate and pressure pulsations. The problem is caused by the mechanical operation of the pump. Basically, the problem occurs when the flow rate is very low or when the syringe size is not appropriate for the selected flow rate. In order to overcome this limitation, one should select thoroughly the volume of the syringe, a small syringe diameter improves the flow rate stability. It is important to take this into account when the expected stability is on the order of magnitude of 0,1 μL/min [9]. An alternative option is using a pressure pump that is free of pulses, as there is not any moving mechanical part connected to the fluid.

> The use of syringe pumps will introduce pulses that would lead to oscillations in the flow rate. The selection of the smaller syringe diameter according to each desirable flow rate is essential to overpass this limitation.

3.4 Remarks

Microfluidics techniques for the fluid-flow characterization have important applications. The selection and applicability of these techniques must be thoroughly examined in order to make the most of their potentialities. When the objective is just a qualitative assessment of the flow field, streak line photography is the most suitable method. However, when one needs to quantify the magnitude of the velocity field, the μ-PIV is the most recommendable technique. Nevertheless, the time comsuption of the latter one is higher due to the data post processing, but with a more valuable information. Additionally, the μ-PIV is an expensive and bulky technique.

The characterization of the pressure drop in a micro device is also convenient even when the measurements are always the most complicated of all of the mentioned techniques. An excellent complement to all this information about the fluid flow is the determination of the stress distribution in polymer solutions that can be obtained using birefringence techniques.

3.5 Summary

In this book chapter the main microfluidic techniques used in the characterization of the complex fluid flows have been reviewed as well as some practical tips that may be help in their use.

#1 A proper design of the microfluidic set up is crucial for obtaining reliable and accurate measurements of the main flow characteristics.

#2 Knowing the potentialities and applicability of the available experimental techniques is crucial for the selection of the most appropriate method for fluid flow characterization.

#3 The applicability of FIB techniques requires the use of birefringent materials introducing a restriction to the fluids being used.

#4 Birefringence measurements provide the stress tensor directly only when the flow is considered to be 2D. Otherwise, for a 3D flow, the birefringence and the orientation angle vary along the light path leading to a cumulative measurement of the optical properties.

#5 In the case of micro channels with a circular cross-section geometry the refractive index of the working fluid should match the refractive index of the micro channel material in order to avoid optical distortions.

#6 The determination of the entrance length when the working fluid is non-Newtonian is not trivial. This distance could be only determined by measuring the velocity profiles along the micro channel.

#7 The rising of air bubbles affects the flow and avoids the obtention of accurate and reliable measurements. It is important to verify that the micro channel is free of air bubbles.

#8 Sedimentation, aggregation and stuck of tracer particles in the channel or in the supply tubes should be avoided by adding specific solutions and by matching the density between particles and the working fluid.

#9 Before performing any flow characterization technique, the focussing of the mid plane must be guaranteed.

#10 The use of syringe pumps will introduce pulses that would lead to oscillations in the flow rate. The selection of the smaller syringe diameter according to each desirable flow rate is essential to overpass this limitation.

References

1. Bagley, E. B. (1957). End corrections in the capillary flow of polyethylene. *Journal of Applied Physics, 28,* 624–627.
2. Campo-Deaño, L., Galindo-Rosales, F. J., Oliveira, M. S. N., Alves, M. A., & Pinho, F. T. (2011). Flow of viscosity Boger fluids through a microfluidic hyperbolic contraction. *Journal of Non-Newtonian Fluid Mechanics, 166,* 1286–1296.
3. Campo-Deaño, L., Dullens, R. P. A., Aarts, D. G. A. L., Pinho, F. T., & Oliveira, M. S. N. (2013). Viscoelasticity of blood and viscoelastic blood analogues for use in polydimethylsiloxane in vitro models of the circulatory system. *Biomicrofluidics, 7,* 034102.
4. Campo-Deaño, L. (2016). Assessing the dynamic performance of microbots in complex fluid flows. *Applied Sciences, 6,* 410.
5. Calejo, D., Pinho, D., Galindo-Rosales, F. J., Lima, R., & Campo-Deaño, L. (2016). Particulate blood analogues reproducing the erythrocytes cell free layer in a microfluidic device containing a hyperbolic contraction. *Micromachines, 7,* 4.
6. Cheung, P., Toda-Peters, K., & Shen, A. Q. (2012). In situ pressure measurement within deformable rectangular polydimethylsiloxane microfluidic devices. *Biomicrofluidics, 6,* 026501.
7. Durst, F., Melling, A., & Whitelaw, J. H. (1981). *Principles and practice of laser-Doppler anemometry* (2nd ed.). London: Academic Press.
8. Eaton, W. P., & Smith, J. H. (1997). Micromachined pressure sensors: review and recent developments. *Smart Materials and Structures, 6,* 530–539.
9. Elveflow http://www.elveflow.com/microfluidic-tutorials/microfluidic-reviews-and-tutorials/ syringe-pumps-and-microfluidics/stability-and-flow-oscillation-of-syringe-pumps-in-microfluidic/ Cited 30 April 2017.
10. Galindo-Rosales, F. J., Alves, M. A., & Oliveira, M. S. N. (2013). Microdevices for extensional rheometry of low viscosity elastic liquids: A review. *Microfluidics and Nanofluidics, 14,* 1–19.
11. Galindo-Rosales, F. J., Oliveira, M. S. N., & Alves, M. A. (2014). Optimized cross-slot microdevice for homogeneous extension. *RSC Advances, 4,* 7799–7804.
12. Haward, S. J., McKinley, G. H., & Shen, A. Q. (2016). Elastic instabilities in planar elongational flow of monodisperse polymer solutions. *Sicentific Reports, 6,* 33029.
13. Larson, R. G., Khan, S. A., & Raju, V. R. (1988). Relaxation of Stress and Birefringence in Polymers of High Molecular Weight. *Journal of Rheology, 32,* 145–161.
14. Martínez-Aranda, S., Galindo-Rosales, F. J., & Campo-Deaño, L. (2016). Complex flow dynamics around 3D microbot prototypes. *Soft Matter, 12,* 2334–2347.
15. Mishra, S., Thakur, A., Redenti, S., & Vazquez, M. (2015). A model microfluidics-based system for the human and mouse retina. *Biomed Microdevices, 17,* 107.
16. Mitsoulis, E. (1998). Numerical simulation of confined flow of polyethylene melts around a cylinder in a planar channel. *Journal of Non-Newtonian Fluid Mechanics, 76,* 327–350.
17. Nam, S.-W., Choi, S., Cheong, Y., Kim, Y.-H., & Park, H.-K. (2015). Evaluation of aneurysm-associated wall shear stress related to morphological variations of circle of Willis using a microfluidic device. *Journal of Biomechanics, 48,* 348–353.
18. Nguyen N.-T., Wereley, S. T. (2006). *Fundamental and applications of microfluidics,* (2nd ed.), Artech House, Boston.
19. Ober, T. J., Haward, S. J., Pipe, C. J., Soulages, J., & McKinley, G. H. (2013). Microfluidic extensional rheometry using a hyperbolic contraction geometry. *Rheologica Acta, 52,* 529–546.
20. Oliveira, M. S. N., Alves, M. A., Pinho, F. T. (2011). Microfluidic flows of viscoelastic fluids. In R. Grigoriev (Ed.), *Transport and mixing in laminar flows: From microfluidics to oceanic currents.* Germany:Wiley-VCH Verlag GmbH & Co. KGaA, Weinheim.
21. Olsen, M. G., & Adrian, R. J. (2000). Out-of-focus effects on particle image visibility and correlation in microscopic particle image velocimetry. *Experiments in Fluids, 29,* S166–S174.
22. Pan, L., & Arratia, P. E. (2013). A high-shear, low Reynolds number microfluidic rheometer. *Microfluidics and Nanofluidics, 14,* 885–894.

23. Pipe, C. J., & McKinley, G. H. (2009). Microfluidic rheometry. *Mechanics Research Communications, 36,* 110–120.
24. Pries, A. R., Secomb, T. W., Gaehtgens, P., & Gross, J. F. (1990). Blood flow in microvascular networks: Experiments and simulation. *Circulation Research, 67,* 826–834.
25. Sinton, D. (2004). Microscale flow visualization. *Microfluidics and Nanofluidics, 1,* 2–21.
26. Sousa, P. C. (2010). *Entry flow of viscoelastic fluids at macro- and micro- scale.* PhD Thesis.
27. Squires, T. M., & Quake, S. R. (2005). Microfluidics: Fluid physics at the nanoliter scale. *Reviews of Modern Physics, 77,* 977.
28. Stone, S. W., Meinhart, C. D., & Wereley, S. T. (2002). A microfluidic-based nanoscope. *Experiments in Fluids, 33,* 613–619.
29. Takahashi, T., & Fuller, G. (1996). Stress tensor measurement using birefringence in oblique transmission. *Rheologica Acta, 35,* 297–302.
30. Wereley, S. T., & Meinhart, C. D. (2005). *Micron-resolution particle image velocimetry, in micro- and nanoscale diagnostic technologies.* New York: Springer-Verlag.
31. Haward, S. J., & Zhou, L. (2015). Viscoelastic flow development in planar microchannels. *Microfluidics and Nanofluidics, 19,* 1123–1137.
32. Zijl, J. L. J. (2001). *Polymers in extensional flow. A comparison of microscopic and macroscopic simulations with birefringence experiments.* PhD Thesis.
33. Zilz, J., Schäfer, C., Wagner, C., Poole, R. J., Alves, M. A., & Lindner, A. (2014). Serpentine channels: micro-rheometers for fluid relaxation times. *Lab on a Chip, 14,* 351.

Chapter 4
Numerical Simulations of Complex Fluid-Flows at Microscale

Alexandre M. Afonso

Abstract This chapter provides to the students an up-to-date overview of the current challenges in the numerical simulations of complex fluid-flows at microscale. At such reduced length scales, phenomena involving viscous forces, reactive flows and surface effects (electrokinetics, slip flows, capillarity, surface tension) may become more important, and eventually overcame the effects of other macro scale predominant forces, such as inertial or gravitational effects. Therefore, numerical methods to simulate complex fluid-flows at microscale need to account with this new forces interplay arrangement. This chapter includes detailed insights into the main differences between macro and micro numerical simulations, presented in a modular view, and containing the synopsis of the theoretical models, numerical methods and simulation tools.

4.1 Introduction

Microfluidic devices are expected to become the next-generation laboratory platforms for diverse areas such as biology, chemistry and medicine [1–4]. There is a large expectation that the miniaturization of fluid circuits will have an impact on chemistry, biology, and energy systems as dramatic as the revolution brought by the integrated circuit, which changed electronic computations, control systems and manufacture beyond recognition, thus permeating to all kinds of consumer products. If that is to happen, progress must occur not only in microfabrication, but also in miniature detection and manipulation devices to allow for the possibility of cheap portable devices performing physical, chemical and biological analysis, amongst others. Microfluidics is expected to be the key growth market for industry in Europe, with an expected \$5 Billion market in 2016 [2]. Considering only the life sciences and the new medical devices areas, the market value is expected to around \$4.5 Billion in 2022 [2].

A.M. Afonso (✉)
Centro de Estudos de Fenómenos de Transporte, Departamento de Engenharia Mecânica,
Faculdade de Engenharia da Universidade do Porto, Porto, Portugal
e-mail: aafonso@fe.up.pt

© Springer International Publishing AG 2018
F.J. Galindo-Rosales (ed.), *Complex Fluid-Flows in Microfluidics*,
DOI 10.1007/978-3-319-59593-1_4

In the context of Newtonian fluid flows, the use of numerical methods that properly deal with phenomena involving low Reynolds (Re) and capillary (Ca) numbers is very limited [5]. At microscales the Reynolds number of interest is usually very low ($Re \ll 1$) providing a challenging numerical problem that must be solved. When working with complex fluids, i.e., with the inclusion of elastic effects, quantified by the Weissenberg number (Wi), increases the level of complexity and complications in the numerical modeling. Therefore, numerical algorithms for microfluidics need to solve fascinating dynamics of the non-Newtonian multiphase flows in microfluidics, combining $Re \ll 1$, $Ca \ll 1$ and $Wi \gg 1$.

The level of complexity also increases with the complexity of the practical applications, as when the equations for the conservation of linear momentum are conjugated with mass and energy conservation equations, or when new body forces are taken into account in order to deal with other physical phenomena. Usually, these new forcing terms become important when they scale with surface area and the fluid transport design strategies are scaled down to microscales, as in microfluidic devices. When energy conservation is accounted, for instance when dealing with heat transfers or reactive flows at microscale (eg. in micro-combustion), the simultaneous conjugated heat transfer from the fluid-phase to the solid boundary needs to be accounted, due to the increase of the surface to bulk ratio and the heat loss through the solid walls.

Therefore, simulations tools for micro-physics problems, specially for those including complex fluids with complex thermal, rheological or superficial behavior (non-Newtonian, viscoelastic fluids, hot gases, melted plastics, etc.) and complex interactions (configurational multiphase changes, temperature or concentration variations, surface tension gradients, electrokinetic body forces, reaction kinetics, combustion, elastic instabilities, slip, etc.) present a significant computational challenge.

This chapter is further divided into three main sections. In Sect. 4.2, the main theoretical concepts in microfluidics are presented and the major differences between macro and micro regimes are addressed. Section 4.3, is reserved for an up-to-date overview of the most important types of numerical methods and simulations tools adapted to microscale flows. Finally, a summary of key tips on numerical simulations of complex fluid-flow at microscales is presented in Sect. 4.4.

4.2 Theoretical Concepts

The fundamental laws in microfluidic systems are the same as in macro systems. In this course we will use well known laws from mechanics, fluid dynamics, electromagnetism, thermodynamics and physical chemistry. The major differences is that the relative importance of different forces change with geometric scaling.

The fundamental laws in microfluidic systems are the same as in macro systems. The major differences is that the relative importance of different forces change with the geometric scaling.

Here, in this brief theoretical introduction, we will assume processes in the Mechanics of continuum media, which involve the transfer of mass, momentum and energy through and with matter. These processes are usually termed Transport Phenomena, and include diffusion processes, heat transport and fluid dynamics [6]. Fluid dynamics can be defined as the science that studies the motion of fluids, and was founded under the classical mechanics conservative axioms derived under the continuum hypothesis (here again, given the earthly thermodynamic standard PVT conditions, the quantum mechanics and general relativity theories are discarded), namely the mass, momentum and energy conservation.

The continuity equation. The conservation of mass in a microscale flow can be described by the continuity equation,

$$\frac{\partial \rho}{\partial t} + \nabla \cdot \rho \mathbf{u} = 0 \tag{4.1}$$

where t is the time, ρ the fluid density and $\mathbf{u}$ is the velocity vector. In many situations, and especially in microscale flows, in which the velocities are much smaller than the velocity of sound, the fluid can be represented as incompressible. This means that ρ is constant in time and space and the continuity equation reduces to

$$\nabla \cdot \mathbf{u} = 0 \tag{4.2}$$

The Navier–Stokes equations/Cauchy equations. The conservation of momentum can be described by the Navier–Stokes equations, or if the body force term is accounted, sometimes called as Cauchy equations. The Navier–Stokes/Cauchy equations for a complex fluid can be expressed as:

$$\rho \left[\frac{\partial \mathbf{u}}{\partial t} + \nabla \cdot \mathbf{uu} \right] = -\nabla p + \nabla \cdot \boldsymbol{\tau} + \eta_s \nabla^2 \mathbf{u} + \mathbf{F} \tag{4.3}$$

p the pressure, η_s the Newtonian solvent viscosity and τ the extra stress contribution from the complex fluids (polymeric viscoelastic fluids). $\mathbf{F}$ are body forces, representing external forces that act throughout the fluid.

At microscales the viscous forces effects increases and the Reynolds number is usually very low (*laminar* flow, $Re \ll 1$). The prevailing mechanism for momentum transport is diffusion, which can impose restrictive limitations on the simulations time-step, since the diffusive characteristic time can be represented as $t_{diff} = \rho L^2/\eta$.

One of the major effects when reducing the geometric scale down to microns, is the increase of the viscous over the inertial forces. One way to measure the interplay between these two forces is the Reynolds number, that measures the importance of the ratio between the forces due to fluid acceleration and due to viscous forces. The Reynolds number is a dimensionless parameter, defined as

$$Re = \frac{\text{inertial forces}}{\text{viscous forces}} = \frac{\rho U^2 L^2}{\eta_s U L} = \frac{\rho U L}{\eta_s} \tag{4.4}$$

where U and L are the characteristic velocity and length scales.

For small Reynolds number flows, also called *laminar* flows, the prevailing mechanism for momentum transport is diffusion. For large Reynolds number flows, also called *turbulent* flows, viscous effects can be neglected except close to walls and elsewhere the dominant mechanism of momentum transport is convection. Additionally, at very high Reynolds numbers laminar flows are unstable to infinitesimal perturbations and the flows become turbulent, but even here viscous effects cannot be neglected close to walls, in the boundary layers.

Complex fluids. Complex fluids, such as viscoelastic fluids, can be described by polymeric constitutive equations. These constitutive equation can be described by defining a set of micro-mechanical elements that represent the chain of a polymer molecule. One of the simplest elements to represent the structure of the polymer molecule is the mass-spring (dumbbell) model, in which two masses are connected by an elastic spring, as observed in Fig. 4.1. The system of dumbbells are subject to flow forces, such as viscous drag and elastic forces due to deformation of the spring, and forces resulting from Brownian motion. These forces can be modeled with the aid of stochastic and Brownian methods, being possible to establish a hydrodynamic

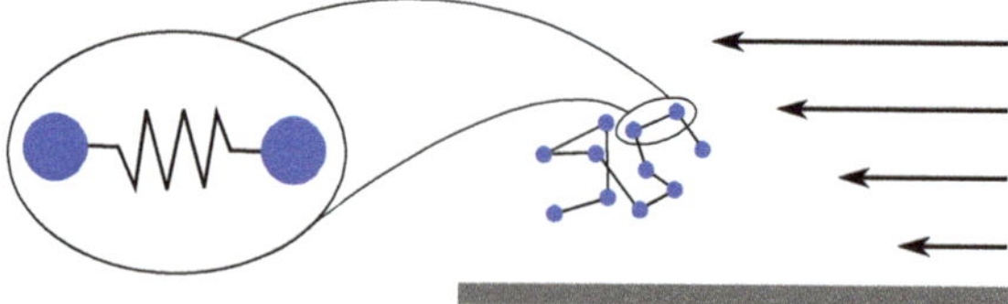

Fig. 4.1 Illustration of the micro-mechanical dumbbell model used to represent a polymer chain. Image courtesy of Rafael Figueiredo (UFO, Brazil)

equation that relates the interaction between the polymer molecules and the solvent [7]. Here, as an example, we will use one of the simplest rheological equation of state, the Oldroyd-B model [7], that can be written as:

$$\tau + \lambda \left[\frac{\partial \tau}{\partial t} + \mathbf{u} \cdot \nabla \tau - \nabla \mathbf{u} \cdot \tau - \nabla \mathbf{u}^T \cdot \tau \right] = 2\eta_p \mathbf{D} \tag{4.5}$$

where η_p is the polymer viscosity coefficient, $\mathbf{D}$ is the rate of deformation tensor and λ is a relaxation time. Since we have introduced a new viscosity coefficient for the polymeric fluid (note that for the Oldroyd-B the viscosity coefficient is constant [7]), we can define a relation between the Newtonian and the polymeric viscosities:

$$\beta = \frac{\text{Newtonian viscosity}}{\text{total viscosity}} = \frac{\eta_s}{\eta_s + \eta_P} = \frac{\eta_s}{\eta_0} \tag{4.6}$$

When simulating viscoelastic fluids flow at microscales, i.e., for high Weissenberg number ($Wi \gg 1$), is recommended to use stabilization methods, namely the log-conformation [8], the square-root [9] or the kernel-conformation [10] methods.

The Weissenberg number,

$$Wi = \frac{\text{relaxation time}}{\text{flow time}} = \frac{\lambda}{t_{flow}} = \frac{\lambda U}{L} = \lambda \dot{\varepsilon} \tag{4.7}$$

is, like the Deborah number, a non-dimensional parameter that can be used to measures the level of elasticity of a fluid. It indicates the degree of anisotropy or orientation generated by the deformation. It can also be represented by the ratio of the polymeric relaxation time and the characteristic time of the flow, t_{flow}.

It is well know, that not so long time ago, the Numerical simulations of viscoelastic fluids at microscale presented a demanding challenge, with the results suffering from a common problem, namely the numerical breakdown for very low Wi values. This problem became known as the High Weissenberg Number Problem, HWNP. This problem on numerical calculations of laminar viscoelastic fluid flows can be better understood by consulting several reference works, particularly [11, 12], but it can shortly be explained as the existence of a limit in the Weissenberg number above which numerical methods diverge. The critical value of the Weissenberg number depends on the flow geometries as well on the constitutive equation, but geometric singularities are usually associated with more severe problems.

In order to deal with the HWNP, several stabilization methods, based on simple matrix transformations of the conformation tensor evolution equation, were proposed, such as the log-conformation [8], the square-root [9] or the kernel-conformation [10] methods.

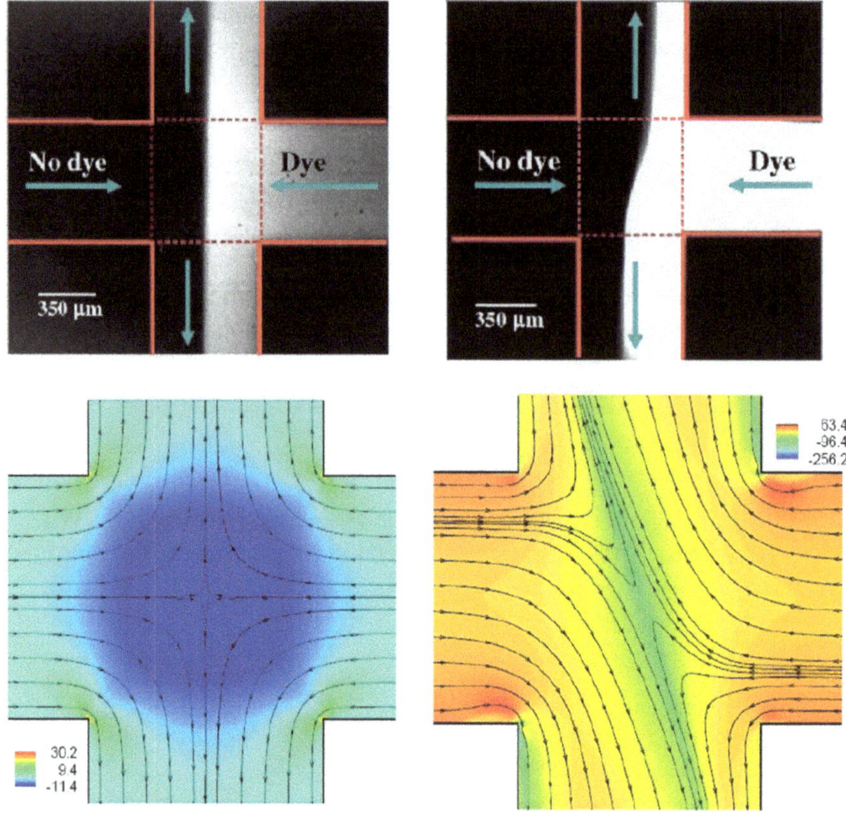

Fig. 4.2 Microfluidic flow in a *cross channel* geometry: (*top*) experimental [14] and (*bottom*) numerical [13] results. Reprinted figure with permission from Ref. [13, 14] Copyright 2017 by the American Physical Society

Poole et al. [13] successfully used a numerical technique with a simple viscoelastic constitutive equation to model the flow at a micro cross slot geometry, which are in good agreement with the steady asymmetry flow obtained by Arratia et al. [14]. The comparison can be observed in Fig. 4.2.

Interfacial effects: Capillarity/Surface tension/multiphase flows. At microscale, since the ratio between area/volume increases, the interfacial effects can be dominant over other forces.

The Capillary number (Ca), that measures the interplay between surface tension to viscous forces, by

$$Ca = \frac{\text{viscous forces}}{\text{surface tension forces}} = \frac{\eta U}{\gamma} \tag{4.8}$$

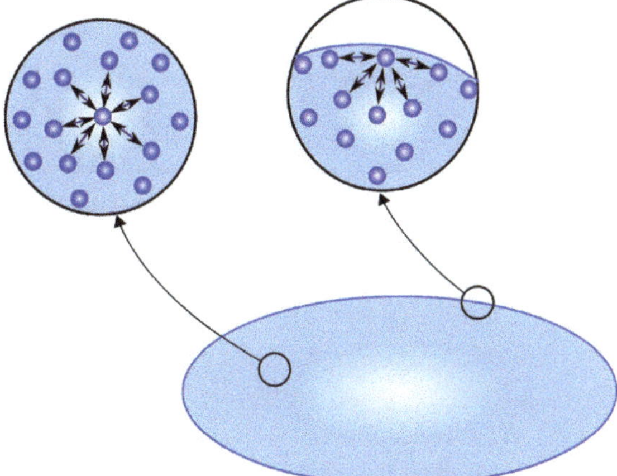

Fig. 4.3 Illustration of the cohesive force in a portion of fluid. Image courtesy of Rafael Figueiredo (UFO, Brazil)

where γ is the surface tension coefficient.

The surface tension is a phenomenon that occurs at the interface between different liquids due to the cohesive force. Cohesion strength is a force of attraction between molecules of the same material. Figure 4.3 presents a illustrative representation of the cohesive force in a portion of fluid. Inside the fluid the molecules are drawn in all directions, and in the balance of forces there is no resulting force. On the other hand, the balance of forces of the interface molecules, results in a force toward the interior of the fluid. This force that pulls the molecules of the interface into the fluid, causes the interface to have an elastic behavior, trying to minimize the interface area. The energy of the interface is the work required to create a new interface. The molecules that are at the interface have higher energy, and are prone to decrease energy by minimizing the interface area.

The scale-down shift of dominating forces, related with the increased dominance of surface forces ($Ca \ll 1$) over inertial forces ($Re \ll 1$), introduces inherent numerical difficulties, such as accurate interface representation. The major challenges in simulating multiphase flows at microscale, is the representation and correct updating of the interface. An additional difficulty arises when the flow is influenced by surface tension due to the accuracy of the curvature calculation.

At very low capillary number flows ($Ca \ll 1$), the inaccuracies on the interface representation, especially when dealing with surface tension models, can originate unbalanced results that can lead into the so-called *parasitic currents*. This currents are artificial vortex-like structure flows in the neighborhood of the interface. To address this problem, please consult the work by Raessi et al. [15]. For a review on numerical methods for multiphase flows at microscale consult the work by Worner [16].

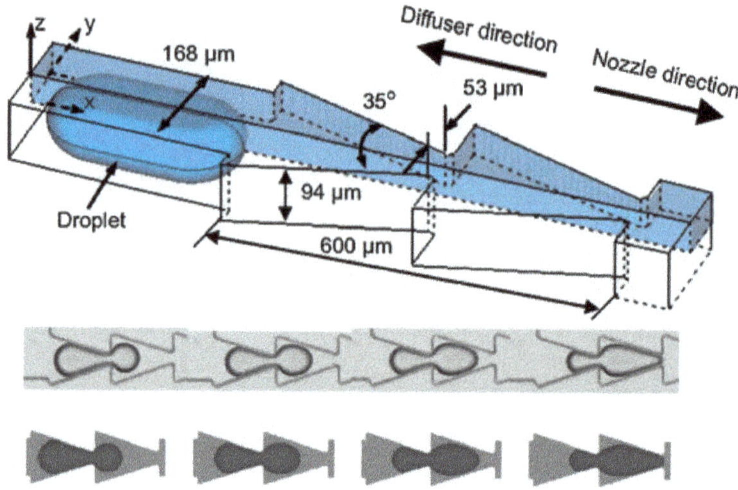

Fig. 4.4 Numerical simulation and experimental results of the motion of a droplet through microfluidic ratchets [17]. Reprinted figure with permission from Ref. [17] Copyright 2017 by the American Physical Society

The body force term in the Navier–Stokes equation, $\mathbf{F}$, related to the surface tension force, can be defined as

$$\mathbf{F} = \gamma \kappa \mathbf{n} \delta_i, \tag{4.9}$$

where $\mathbf{n}$ is the normal vector to the interface, κ is the interface curvature and δ_i is the δ-function at the interface. Numerically, the representation of the interface between fluids, can be obtained by solving the transport equation for the volume fraction, α:

$$\frac{\partial \alpha}{\partial t} + \nabla \cdot (\alpha \mathbf{u}) = 0. \tag{4.10}$$

Finally, the properties of the fluid can be obtained at each phase (1 and 2) as

$$\chi = \alpha \chi_1 + (1 - \alpha) \chi_2, \tag{4.11}$$

where χ can be any property of the fluid.

Figure 4.4 presents the numerical simulation and experimental results of the motion of a droplet through microfluidic ratchets [17].

Electrokinetics Phenomena. In Electrokinetic phenomena (also known as electro-fluid-dynamics or electro-hydrodynamics) such as electro-osmotic flow, the Navier–Stokes equations are coupled with the Maxwell's and the Poisson–Nernst–Planck equation.

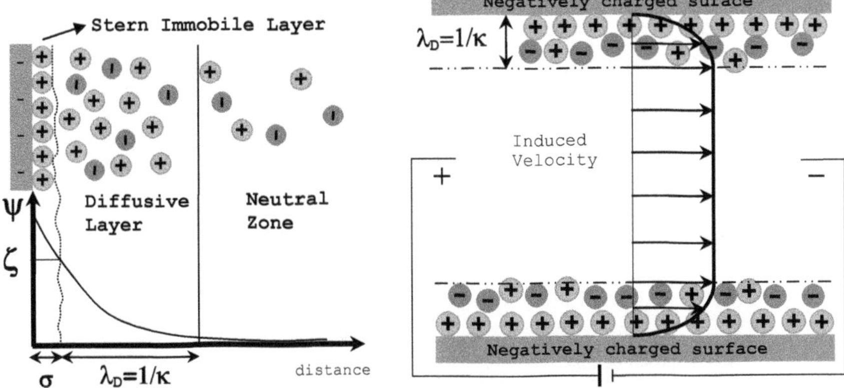

Fig. 4.5 Schematic of the ions and potential distribution near a charged wall

Electro-osmotic flow is induced by an additional body force term, the applied external electric field. A natural arrangement of ions occurs when a polar fluid interacts a dielectric charged wall. In Fig. 4.5, the negatively charged wall attracts the fluid's positive ions, forming several layers of positively charged fluid near the walls. The first layer is called Stern immobile layer, σ, and is composed of counter-ions, while the second layer is called diffusive layer, λ_D, where the ions have the ability to more around freely. The different concentrations of counter-ions and co-ions leads to the creation of a varying potential field, ψ, as observed in Fig. 4.5. These layers are also known as the Electric double Layer (EDL). When an external electric field is applied, a body force will act on the counter-ions of the EDL. The movement of this counter-ions will drag the neutral conducting fluid above by viscous forces, as observed in Fig. 4.5.

For electro-osmosis, the coupling body term, **F**—see Eq. (4.3), can be defined as

$$\mathbf{F} = ez\left(n^+ - n^-\right)\nabla(\phi + \psi) \tag{4.12}$$

where ϕ represents the external applied electric field, e is the elementary electric charge and n^+ and n^- are the cation and anion concentrations, respectively.

The Poisson–Nernst–Planck equations, defined as:

$$\frac{\partial n^\pm}{\partial t} + \mathbf{u}\cdot\nabla n^\pm = \nabla\cdot\left(D^\pm\nabla n^\pm\right) \pm \nabla\cdot\left[D^\pm n^\pm \frac{ez}{k_B T}\nabla\left(\phi + \psi\right)\right] \tag{4.13}$$

are a conservation of species equation used to describe the motion of n^+ and n^- in a fluid medium, in which the diffusing species are also in motion with respect to the fluid by electrostatic forces. In Eq. (4.13), D^+ and D^- are the diffusion coefficients for n^+ and n^-, respectively.

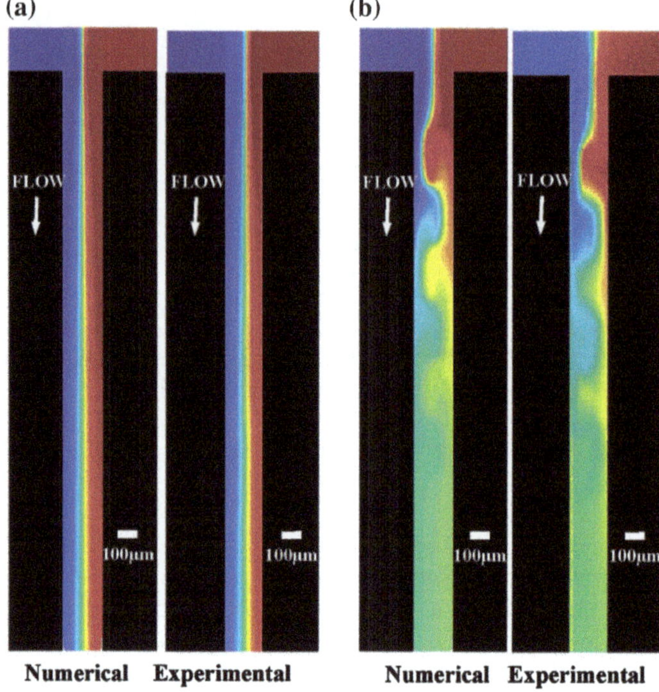

Fig. 4.6 Numerical simulations and experimental results for a microscale flow with enhanced electrokinetic mixing [18]. Reproduced from Ref. [18] with permission of the American Chemical Society

The comparison between the numerical simulations and experimental results for a microscale flow with enhanced electrokinetic mixing [18], can be observed in Fig. 4.6

Heat Transfer and Reactive flows. Heat transfer and reactive flows are also governed by the mass and momentum conservation, along with the corresponding energy balance. Microscale flow of a reactive fluid, such as in micro-combustion with homogeneous reactions, have been studied due to the potential applications in micro devices systems as a power supply, because of the high energy density when compared with the typical batteries.

Nevertheless, it is known that heat release impacts at small scales result in different behaviors when compared to conventional macro and mesoscales, leading to several practical limitations in its design and performance. In order to circumvent the issue of energy losses at microscale (and also for meso and macroscales), some innovative systems were devised (some examples of these applications can be observed in Fig. 4.7), such as the Swiss-roll burner [19], in which the heat released by the exothermic kinetic reaction is recirculated inside the burner. This heat recirculation will pre-heat the inlet premixed gases, increasing the flame stabilizations.

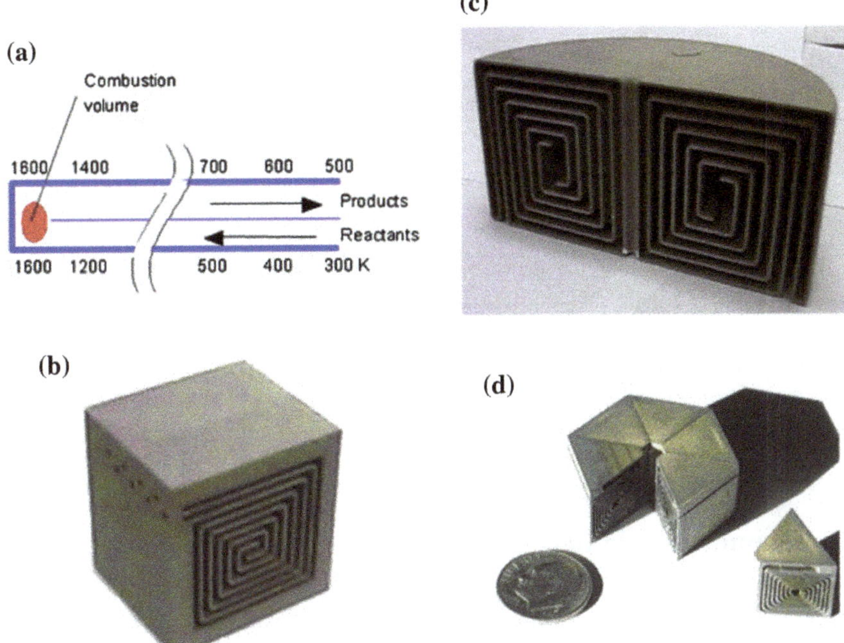

Fig. 4.7 Illustrations of heat recirculating micro-burners [19]. Reprinted from publication Ref. [19] Copyright 2017, with permission from Elsevier

The equations that describe the microscale flow of a reactive fluid are the species conservation:

$$\frac{\partial}{\partial t}(\rho Y_k) + \nabla(\rho Y_k \mathbf{u}) = -\nabla(\rho Y_k \mathbf{V_k}) + \dot{R}_k \quad k = 1, \ldots, N_C \tag{4.14}$$

and the energy conservation

$$\rho C_P \frac{\partial T}{\partial t} + \rho C_P \mathbf{u} \nabla T = -\nabla \mathbf{q} - \rho \sum_{k=1}^{N_C} C_{P,k} Y_k V_k - \sum_{k=1}^{N_C} h_k \dot{\Omega}_k \tag{4.15}$$

where N_C is the total number of species, T is the temperature, Y_k the mass fraction, $\mathbf{V_k}$ is the diffusion velocity, $\dot{R}_k$ the formation rate, h_k the individual species enthalpy and C_i are the specific heat. The heat flux vector, $\mathbf{q}$, is defined as:

$$\mathbf{q} = -\lambda \nabla T + \mathbf{q}_{rad} \tag{4.16}$$

When energy conservation is accounted, for instance when dealing with heat transfers or reactive flows at microscale (eg. in micro-combustion), the simultaneous conjugated heat transfer between the gas-phase and the combustion chamber walls needs to be accounted, due to the increase of the surface to bulk ratio and the heat loss through the solid walls.

When simulating reactive flows or flows with heat transfer at microscales, in order to obtain realistic results, the simultaneous conjugated heat transfer between the fluid-phase and the solid boundary needs to be accounted, due to the increase of the surface to bulk ratio and the heat loss through the solid walls. This effect can be observed in Fig. 4.8, that presents the numerical simulation and experimental results obtained in a micro-jet hydrogen flames [20].

4.3 Numerical Methods

Overview of numerical methods. Even for Newtonian and non-complex fluids, classical fluid dynamics is rich in complex nonlinear problems, most of them defying the goal of obtaining exact solutions. Somewhat surprisingly, even when analytical techniques are applied to the Navier–Stokes equations, the mathematicians have not yet been able to prove that the solution always exist in generic three dimensional problems, and when the solution does exist the smoothness could be questionable, i.e., the solution could contain any singularity or discontinuity. Due to restrictions on the purely analytical approach to solve the full complexity of the flow dynamic problems, scientists and especially engineers often rely on modern computational tools and particularly on Computational Fluid Dynamics (CFD) to calculate the flow in many geometries.

Although in the theoretical introduction, Sect. 4.2, we focused mainly in the classical mechanics conservative axioms derived under the *continuum hypothesis*, there are other possible approaches, especially when some characteristic geometric scale are below the *continuum hypothesis*, such as the *sub-continuum* models: micro and mesoscopic methods. A simple way to evaluate if the *continuum hypothesis* is valid, is to estimate the Knudsen number,

$$Kn = \frac{\text{molecular mean free} - \text{path}}{\text{system length}} = \frac{\lambda_M}{d_h} \tag{4.17}$$

which represents a geometric ratio. i.e., the ratio between the molecular mean free path and the system lengths.

Figure 4.9, along with Tables 4.1 and 4.2, present a possible classification for the numerical methods, based on the fluid Knudsen number, devising three possible main numerical paths: macroscopic, mesoscopic and microscopic methods.

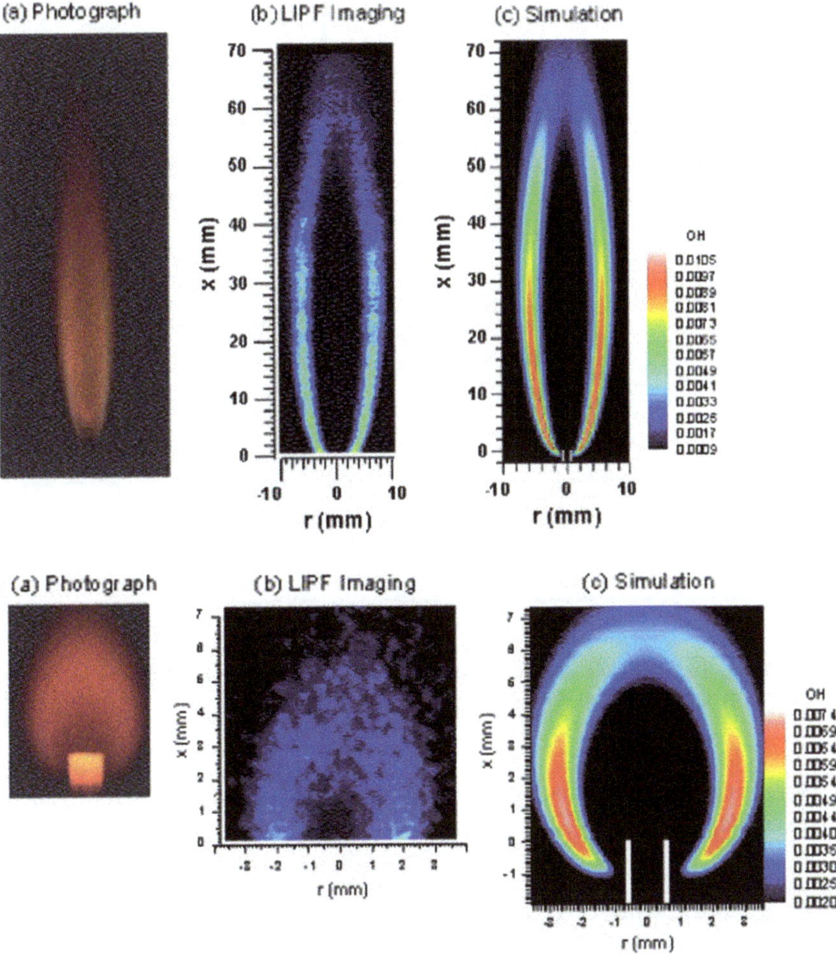

Fig. 4.8 Numerical simulation and experimental results obtained in a micro-jet hydrogen flames [20]: (*top*) higher *Re* with moderate wall effects on the flame and (*bottom*) lower *Re* case with significant wall effects on the flame. Reprinted from publication Ref. [20] Copyright 2017, with permission from Elsevier

In microscopy models, such as Molecular dynamics (MD) and direct simulation Monte Carlo (DSMC), molecules are introduced into a volume of fluid and the corresponding collective forces for each molecule are solved for all molecules. The advantages of this microscopic approaches is that they can be applied to the majority of complex fluids and geometries. The disadvantages is that, when applied to microscale, they may computational inefficient (due to computation load) and the fact that the maximum length scale is close to the lower limit of microscale liquid flows. DSMC is more efficient than MD, with a same level of geometric treatment of

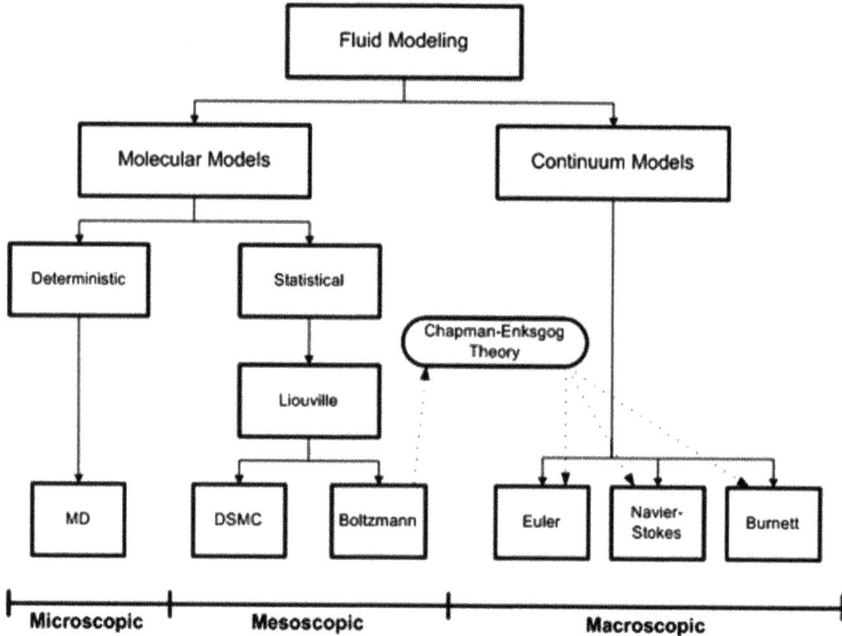

Fig. 4.9 Overview of possible numeric models for fluid dynamics [21]. Reprinted from publication Ref. [21] Copyright 2017, with permission from Elsevier

other macroscopic models, but is not efficient for liquid or complex liquids (is more adequate for gases).

> Molecular dynamics (MD) and direct simulation Monte Carlo (DSMC) are quite flexible (work for complex fluids and geometries), but may be computational inefficient for microscale flows.

In the Mesoscopic methods, particles are introduced into the flow and the governing equations for a group of particles are solved as they evolve inside the flow domain. Examples are the Smoothed Particle Hydrodynamics (SPH, [22]) or the Lattice Boltzmann methods (LBM, [23]), based on the combination of microscopic models and mesoscopic kinetic equations.

Macroscopic methods, rely on the numerical solution of the physical laws introduced in Sect. 4.2, i.e., models in which the *continuum hypothesis* is valid (low Knudsen number $Kn \ll 10^{-3}$). This methods are further explorer in this section.

Computational Fluid Dynamics for macroscopic models, is today a well established and growing scientific branch of fluid dynamics, in which a new level of abstraction and simplification is introduced, with the motto *divide to conquer* being an easy way to explain it. A very schematic view on how the solution is determined

Table 4.1 Recommendation of numerical methods based on the mean molecular free-path (Kn) for gases [21][a]. Reprinted from publication Ref. [21] Copyright 2017, with permission from Elsevier

Kn range	Range description	Recommended models	References
$\rightarrow 0$	Neglect of diffusion	Euler equations	[24]
$\ll 10^{-3}$	Continuum (no slip)	Navier–Stokes equation	[24]
$10^{-3} \ll Kn \ll 10^{-1}$	Continuum with slip	Burnett equations[b]	[25]
$10^{-1} \ll Kn \ll 10$	Transition to molecular flow	Direct Simulation Monte Carlo	[25]
$Kn > 10$	Free-molecule flow	Lattice Boltzmann	[24]
Kn very large	Extreme range	Molecular Dynamics	[24]

[a]This range was collected from Gad-el-Hak [24]; [b]Burnett equations are a high order expansion equations of the Chapman–Enskog equations

Table 4.2 Recommendation of numerical methods based on the mean molecular free-path (Kn) for liquids [21][a]. Reprinted from publication Ref. [21] Copyright 2017, with permission from Elsevier

Kn range	Range description	Recommended models	References
$Kn \ll 10^{-1}$	Continuum	Navier–Stokes equations	[21]
$Kn \gg 10^{-1}$	Sub-continuum or molecular	Meso and Microscopic methods	[21]
Kn very large	Extreme range	Molecular dynamics	[24]

[a]This range was collected from Rosa et al. [21]

in a simple flow problem is the following: the physical domain of the flow dynamic problem is *divided* in very small sub-domains, meaning that the physical domain is *discretized* into a so-called computational grid consisting of very small computational cells; the governing transport equations are also reformulated into a discrete formulation at each computational cell, including the boundary regions, following some simplifications; the solution of the resulting system of discrete equations is achieved via an iterative algorithm, solved with the help of a computer, where the inherent non-linearities of the algebraic equations are dealt with adequately. Obviously, the smaller the sub-domains the better the level of the approximation to the *true* solution of the flow dynamic problem, and in order to achieve the invariance of the computational results with respect to the temporal and spacial discretization, some computational grid and time refinements are needed. The most usual types of numerical procedures to reformulate the governing transport equations in a flow dynamic problem, in which the *continuum hypothesis* is valid, are:

Finite difference method   (FDM, [26]);
Finite volume method   (FVM, [27]);
Finite element method   (FEM, [28]);
Boundary element method   (BEM, [29]).

The finite volume method (FVM) is a numerical method well suited for various types of conservation laws. In this method the conservativeness is explicitly enforced, in sharp contrast to Finite Difference Methods (FDM) and Finite Element Methods (FEM).

> In the FVM conservativeness is explicitly enforced, in sharp contrast to FDM and FEM. A disadvantage of this methods is the appearance of false diffusion when low order numerical methods are used.

The method consists of dividing the domain into control volumes, integrate the differential equation over the control volume and apply the divergence theorem. The derivative terms are evaluated with discrete values at the center of the control volume and this results in a set of linear algebraic equations: one for each control volume. At the end we can solve the system of equations iteratively (the most used procedure in CFD) or simultaneously. There are also some recent procedures that combine some of the above methods, such the hybrid FEM/FVM method [30] or the spectral element method [31], in which a high order finite element method is combined with spectral techniques.

> Lagrangian and ALE methods usually represents the interface accurately, but are rather complicated to implement. Eulerian methods are faster, but suffer from lack of precision in the representation interface.

To simulate multiphase flows at microscale, there are several methods that can be used. These continuum methods for interface representation can be either Lagrangian, Eulerian or an hybrid of Lagrangian-Eulerian methodologies, such the Arbitrary Lagrangian-Eulerian method, ALE. An overview of possible continuum methods for interface representation [16], can be observed in Fig. 4.10. Lagrangian and ALE methods usually represents the interface accurately, but are rather complicated to implement, because of the large distortions mesh involve fluid flow [32, 33]. Eulerian methods usually follow similar laws of Eqs. (4.9)–(4.11) described in the theoretical section. These methods offer lower computational times, but present lower precision in the representation interface, which has to be immersed in the fixed grid. Most of the Eulerian method can not be applied straightforwardly to microscale flow, due to this lack of precision.

Additionally, time-step restriction is a major concern. When performing simulation for high $Wi \gg 1$ and low $Re \ll 1$, the use of explicit methods can be problematic. In this situation the parabolic stability restriction implies that the time step is proportional to the Reynolds number. A possible solution is the use of implicit time methods for the diffusive terms while using explicit methods for the convective terms.

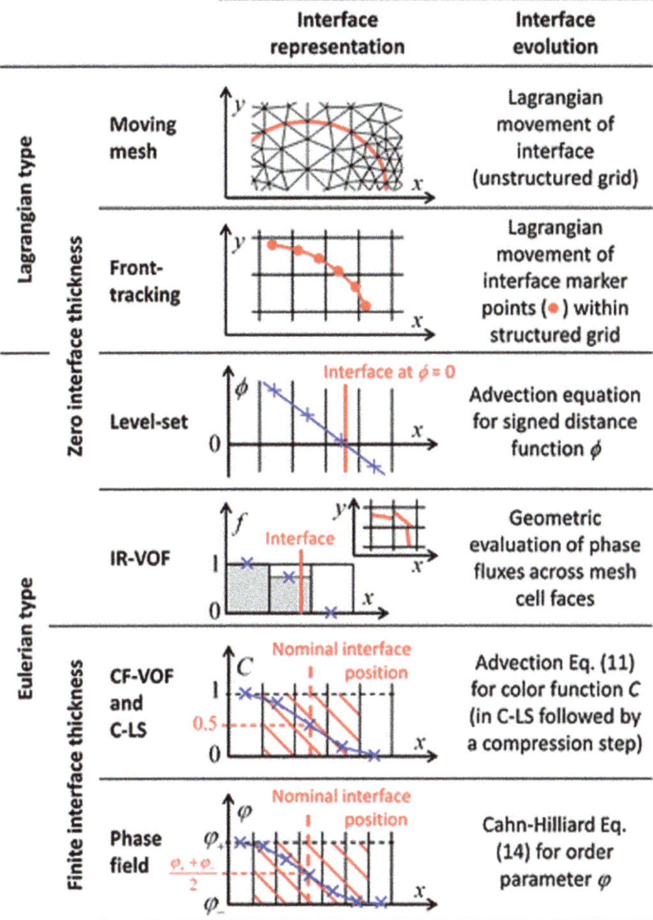

Fig. 4.10 Overview of possible continuum methods for interface representation [16]. Reproduced from Ref. [16] with permission of Springer

When performing simulation for high $Wi \gg 1$ and low $Re \ll 1$, the use of explicit methods can be problematic. In this situation the parabolic stability restriction implies that the time step is proportional to the Reynolds number. A possible solution is the use of implicit time methods for the diffusive terms while using explicit methods for the convective terms.

Overview of simulation tools for microscale simulations. The recommendation of the best simulation tool is not a easy, or even, objective task. All simulations tools have advantages and disadvantages, and the correct choice will depend on several parameters, such as project budget, final motivation (commercial or academic), project duration, time to deliver results, available computational power, etc.

Even between the professional users, this choice is not consensual. A recent survey[1] to CFD users about their software preferences, shown that commercial solutions (Fluent® and Star-CCM+®) remain the dominant packages, although the third player is a open-source tool: OpenFOAM®. One of the main conclusions of this study, is that the actual players tendency is to target to niche markets with specialized functionalities. Here, we will perform this targeting exercise with a focus on simulation tools for complex fluid-flows at microscale.

Open-source:

- OpenFOAM®[2]: is a general purpose open-source CFD code, based on both the finite volume method and molecular dynamics.
- RheoTool[3]: is an open-source toolbox based on OpenFOAM® to simulate the flow of Generalized Newtonian Fluids (GNF) and viscoelastic fluids [34]. It can simulate viscoelastic flows with stabilization methods, multiphase and electrokinetics (in the near future).
- Elmer[4]: is an open source multiphysical simulation software, based on finite element method. It can simulate viscoelastic, multiphase and electrokinetics. In our best knowledge, no stabilization methods for viscoelastic fluids is included.
- FEniCS[5]: An open-source package for computational mathematical modeling. Has some functionality to solve Navier-Stokes, viscoelastic, multiphase and electrokinetics flows. To our best knowledge, no stabilization methods for viscoelastic fluids is included.
- Palabos[6]: is an open-source CFD solver based on the lattice Boltzmann method. It can solve incompressible, isothermal flow, multiphase, free-surface, convective heat transfer and fluid-particle interaction. To our best knowledge, no viscoelastic fluids methods, neither combustion is included.
- laminarSMOKE[7]: CFD solver (based on OpenFOAM®) for laminar reacting flow with detailed kinetic mechanisms. For micro-combustion simulations, a conjugated heat transfer version is needed.
- many others.[8]

Commercial:

- Ansys® Workbench (Fluent®/CFX®): simulation software for fluid flow, turbulence, heat transfer, and reactions for industrial applications. It is based on the

---

[1] www.resolvedanalytics.com/theflux/comparing-popular-cfd-software-packages.

[2] www.openfoam.com.

[3] http://github.com/fppimenta/rheoTool.

[4] www.csc.fi/web/elmer.

[5] http://fenicsproject.org.

[6] www.palabos.org.

[7] http://github.com/acuoci/laminarSMOKE.

[8] An extended list, with over 200 CFD related software packages, can be found here www.cfd-online.com/Wiki/Codes.

finite volume method. For viscoelastic simulations use PolyFlow®. To our best knowledge, no stabilization methods for viscoelastic fluids is included.

- COMSOL®: is a finite element analysis, solver and simulation software package for several physics and engineering applications. It can simulate viscoelastic flows with stabilization methods,[9] multiphase and electrokinetics.
- STAR-CCM+®: is an all-in-one solution for multidisciplinary engineering simulation. To our best knowledge, no stabilization methods for viscoelastic fluids is included.
- many others.

> The tendency described by the actual CFD professionals is to target to niche markets with specialized functionalities. In the author opinion, open-source simulation tools (namely OpenFOAM® and RheoTool [34]) are good candidates to perform simulations for complex fluid-flows at microscale.

4.4   Resume

In this chapter we have covered the detailed differences between macro and micro numerical simulations, presenting the suitable theoretical models, numerical methods and simulation tools. A list of the major key tips on numerical simulations of complex fluid-flow at microscales can be summarize as:

#1 The fundamental laws in microfluidic systems are the same as in macro systems. The major differences is that the relative importance of different forces change with geometric scaling.

#2 At microscales the viscous forces effects increases and the Reynolds number is usually very low (*laminar* flow, $Re \ll 1$). The prevailing mechanism for momentum transport is diffusion, which can impose restrictive limitations on the simulations time-step, since the diffusive characteristic time can be represented as $t_{diff} = \rho L^2/\eta$.

#3 When simulating viscoelastic fluids flow at microscales, i.e., for high Weissenberg number ($Wi \gg 1$), is recommended to use stabilization methods, namely the log-conformation [8], the square-root [9] or the kernel-conformation [10] methods.

#4 When performing simulation for high $Wi \gg 1$ and low $Re \ll 1$, the use of explicit methods can be problematic. In this situation the parabolic stability restriction implies that the time step is proportional to the Reynolds number. A possible solution is the use of implicit time methods for the diffusive terms while using explicit methods for the convective terms.

---

[9] http://arxiv.org/pdf/1508.01041v2.pdf.

#5 When dealing with reactive flows at microscale (eg. in micro-combustion), the simultaneous conjugated heat transfer between the gas-phase and the combustion chamber walls needs to be accounted, due to the increase of the surface to volume ratio and the heat loss through the solid walls.

#6 Lagrangian and ALE methods usually represents the interface accurately, but are rather complicated to implement. Eulerian methods are faster, but suffer from lack of precision in the representation interface.

#7 At very low capillary number flows ($Ca \ll 1$), the inaccuracies on the interface representation, especially when dealing with surface tension models, can originate unbalanced results that can lead into the so-called *parasitic currents*. This currents are artificial vortex-like structure flows in the neighborhood of the interface. To address this problem, please consult the work by Raessi et al. [15]. For a review on numerical methods for multiphase flows at microscale consult the work by Worner [16].

#8 Molecular dynamics (MD) and direct simulation Monte Carlo (DSMC) are quite flexible (work for complex fluids and geometries), but may be computational inefficient for microscale flows.

#9 In the finite volume method (FVM) conservativeness is explicitly enforced, in sharp contrast to Finite Difference Methods (FDM) and Finite Element Methods (FEM). A disadvantage of this methods is the appearance of false diffusion when low order numerical methods are used.

#10 The tendency described by the actual CFD professionals is to target to niche markets with specialized functionalities. In the author opinion, open-source simulation tools (namely OpenFOAM® and RheoTool [34]) are good candidates to perform simulations for complex fluid-flows at microscale.

Acknowledgements A.M. Afonso acknowledges the funding by FCT, COMPETE and FEDER through Project No. PTDC/EMS-ENE/3362/ 2014. The author would also like to thank Prof. Rafael Figueiredo from Univeridade Federal de Uberlândia, Brazil, for the assistance and helpfull discussion regarding Figs. 4.1 and 4.3

References

1. Whitesides, G. M. (2006). The origins and the future of microfluidics. *Nature, 442*, 368–373.
2. In: KARIM Foresighting on microinjection moulding technology (2014) KARIM. knowledge acceleration and responsible innovation meta-network. Available via DIALOG. http://www.karimnetwork.com/wp-content/uploads/2014/04/KARIM-foresight-report-4.pdf. Cited 17 Jan 2017.
3. Pennathur, S. (2008). Flow control in microfluidics: are the workhorse flows adequate? *Lab on a Chip, 8*, 383–387.
4. Dendukuri, D., & Doyle, P. S. (2009). The synthesis and assembly of polymeric microparticles using microfluidics. *Advanced Materials, 21*, 4071–4086.
5. Hoang, D. A., van Steijn, V., Portela, L. M., Kreutzer, M. T., & Kleijn, C. R. (2013). Benchmark numerical simulations of segmented two-phase flows in microchannels using the volume of fluid method. *Computers & Fluids, 86*, 28–36.

6. Bird, R. B., Stewart, W. E., & Lightfoot, E. N. (2002). *Transport phenomena* (2nd ed.). New York, NY: Wiley.

7. Bird, R. B., Armstrong, R. C., & Hassager, O. (1987). *Dynamics of polymeric liquids: fluid mechanics* (2nd ed.). New York, NY: Wiley.

8. Fattal, R., & Kupferman, R. (2004). Constitutive laws for the matrix-logarithm of the conformation tensor. *Journal of Non-Newtonian Fluid Mechanics, 123*, 281–285.

9. Balci, N., Thomases, B., Renardy, M., & Doering, C. R. (2011). Symmetric factorization of the conformation tensor in viscoelastic fluid models. *Journal of Non-Newtonian Fluid Mechanics*, 166–11:546–553

10. Afonso, A. M., Pinho, F. T., & ALves, M. A. (2012). The kernel-conformation constitutive laws. *Journal of Non-Newtonian Fluid Mechanics, 167–168*, 30–37.

11. Owens, R. G., Phillips, T. N. (2002). Computational rheology. World Scientific.

12. Keunings, R. (1986). On the high Weissenberg number problem. *Journal of Non-Newtonian Fluid Mechanics, 20*, 209–226.

13. Poole, R. J., Alves, M. A., & Oliveira, P. J. (2007). Purely elastic flow asymmetries. *Physical Review Letters, 99*(16), 164503.

14. Arratia, P. E., Thomas, C. C., Durian, D. J., & Gollub, J. P. (2006). Elastic instabilities of polymer solutions in cross-channel flow. *Physical Revision Letters, 96*(14), 144502.

15. Raessi, M., Bussmann, M., & Mostaghimi, J. (2009). A semi-implicit finite volume implementation of the CSF method for treating surface tension in interfacial flows. *International Journal for Numerical Methods in Fluids, 59*, 1093–1110.

16. Wrner, M. (2012). Numerical modeling of multiphase flows in microfluidics and micro process engineering: a review of methods and applications. *Micro Nano, 12*(6), 841–86.

17. Liu, J., Yap, Y. F., & Nguyen, N. T. (2009). Motion of a droplet through microfluidic ratchets. *Physical Review E, 80*(4), 046319.

18. Biddiss, E., Erickson, D., & Li, D. (2004). Heterogeneous surface charge enhanced micromixing for electrokinetic flows. *Analytical Chemistry, 76*, 3208–3213.

19. Ju, Y., & Maruta, K. (2011). Microscale combustion: Technology development and fundamental research. *Progress in Energy and Combustion Science, 37*(6), 669–715.

20. Cheng, T. S., Wu, C. Y., Chen, C. P., Li, Y. H., Chao, Y. C., Yuan, T., et al. (2006). Detailed measurement and assessment of laminar hydrogen jet diffusion flames. *Comb Flame, 146*(1), 268–282.

21. Rosa, P., Karayiannis, T. G., & Collins, M. W. (2009). Single-phase heat transfer in microchannels: The importance of scaling effects. *Applied Thermal Engineering, 29*, 3447–3468.

22. Monaghan, J. J. (1988). An introduction to SPH. *Computer Physics Communications, 48*, 88–96.

23. McNamara, G. R., & Zanetti, G. (1988). Use of the Boltzmann equation to simulate lattice-gas automata. *Physical Review Letters, 61*, 2332–2335.

24. Gad-el-Hak, M. (1999). The fluid mechanics of microdevices: The Freeman scholar lecture. *Journal of Fluids Engineering, 121*, 5–33.

25. Xue, H., Ji, H., & Shu, C. (2003). Prediction of flow and heat transfer characteristics in micro couette flows. *Microscale Thermophysical Engineering, 7*, 51–68.

26. Roache, P. J. (1972). *Computational Fluid Dynamics*. Denver, Colorado: Hermosa Publishers.

27. Patankar, S. V. (1980). Numerical heat transfer and fluid flow. Hemisphere Publishing Corporation.

28. Zienkiewicz, O. C., Taylor R. L. (1989). The finite element method: basic formulation and linear problems. McGraw-Hill College.

29. Becker, A. A. (1992). The boundary element method in engineering. London: McGraw-Hill.

30. Sato, T., & Richardson, S. M. (1994). Explicit numerical simulation of time- dependent viscoelastic flow problems by a finite element/finite volume method. *Journal of Non-Newtonian Fluid Mechanics, 51*, 249–275.

31. Patera, A. T. (1984). A spectral element method for fluid dynamics: laminar flow in a channel expansion. *Journal of Computational Physics, 54*, 468–488.

32. Quan, S. (2011). Simulations of multiphase flows with multiple length scales using moving mesh interface tracking with adaptive meshing. *Journal of Computational Physics, 230*, 5430–5448.
33. Montefuscolo, F., Sousa, F. S., & Buscaglia, G. C. (2014). High-order ALE schemes for incompressible capillary flows. *Journal of Computational Physics, 278*, 133–147.
34. Pimenta, F., & Alves, M. A. (2017). Stabilization of an open-source finite-volume solver for viscoelastic fluid flows. *Journal of Non-Newtonian Fluid Mechanics, 239*, 85–104.

Chapter 5
Numerical Optimization in Microfluidics

Kristian Ejlebjerg Jensen

Abstract Numerical modelling can illuminate the working mechanism and limitations of microfluidic devices. Such insights are useful in their own right, but one can take advantage of numerical modelling in a systematic way using numerical optimization. In this chapter we will discuss when and how numerical optimization is best used.

5.1 Introduction

Within complex fluid-flow, the use of numerical optimization is still a rarity, and thus it is only the most obvious applications, such as [2, 9] rectifiers and cross-slot geometries [7, 10], that have been the subject of numerical optimization.

Numerical modelling of complex fluids is often a challenging task in itself. This goes for viscoelastic fluids in particular, but even generalized-Newtonian models can cause numerical instabilities. Numerical optimization builds on top of modelling, so in order for it to be successful, the model has to be robust towards large variations in the design. I.e. the numerical model is your foundation, so a critical prerequisite for performing numerical optimization is that you

verify the robustness of your numerical model and never perform optimization outside its limits—the optimization result should be independent of the numerical discretization.

Ideally, you should also be able to rate the accuracy of your mathematical model, but you might be able to benefit from an optimization based on an inaccurate model, if you have a lot of design freedom. This is due to the fact that such an optimization has the potential to identify new working mechanisms for your design. Conversely, it is

K.E. Jensen (✉)
Department of Micro- and Nanotechnology, Technical University of Denmark,
Ørsteds Plads, 2800 Kongens Lyngby, Denmark
e-mail: krej@dtu.dk

© Springer International Publishing AG 2018
F.J. Galindo-Rosales (ed.), *Complex Fluid-Flows in Microfluidics*,
DOI 10.1007/978-3-319-59593-1_5

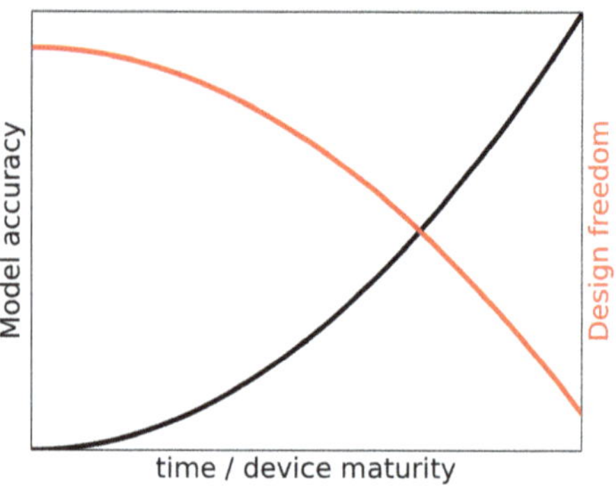

Fig. 5.1 Large design freedom and low model accuracy typically characterizes the early phase of device development, while the opposite is true for later stages. One can thus benefit from numerical optimization throughout the development process, but the potential gain decreases with design freedom

also true that you can benefit from an optimization with little design freedom, if the model is very accurate. The potential gain will be smaller and likely of an evolutionary character, but the probability that the optimization will actually be implemented is high. Note, however, that

> you are unlikely to benefit from an optimization, if you have an inaccurate model and little design freedom.

Such an optimization will result in a small perturbation to the existing design, so it will not identify a new working mechanism and the optimality of the perturbation cannot be trusted due to the inaccuracy of the model. Although early optimization is preferable, it is also worthwhile to note that the understanding of parameters and flow effects increase as a device is developed, which improves the accuracy of the model, see Fig. 5.1.

5.2 Variables

An optimization problem consists of a model, variables, constraints and an objective function. The objective function is used to rate the design and by convention it should be minimized. This is achieved by changing the variables, while respecting the constraints (and the model).

The variables often describe the design, but in principle they can be anything, so the viscosity or other material parameters can also be allowed to vary. Ultimately the choice of variables depends on what you are able to realize, so this choice is strongly tied to the degree of design freedom, whether it be manufacturing constraints or the parameter regime within which the model validity has been verified.

Most optimizers work best, when there are no constraints associated with the variables, which is rarely possible. The second best option is to have a formulation with lower and upper bounds (*box constraints*), so

> try to select the design variables of your problem such that box constraints can
> be used.

To achieve this, one might have to reformulate the variables. In example one could consider the flow past a sphere in a tube with radii r and R, respectively. If the sphere is to fit in the tube, one might impose

$$0 < R < R_{max} \quad \text{and} \quad 0 < r < R$$

as constraints, but the last inequality is not a box constraint, so it is better not to use r as a design variable. Instead one can have a design variables a such that

$$r = aR, \quad \text{where} \quad 0 < a < 1$$

This way box constraints can be used to ensure that the sphere fits in the tube.

5.3 Simultaneous Analysis and Design (SAND)

There are two types of optimization techniques:

#1 The nested formulation, where one alternates between computing the physical variables for the current design and updating the design variables based on the current result of the numerical model.

#2 Simultaneous analysis and design (SAND), where there is no distinction between the design variables and the physical variables. This means that the governing equations are treated as constraints.

The SAND approach should in theory be able to converge much faster and this has also been demonstrated for problems within fluid dynamics [5], but the method cannot guarantee improvement on an initial design and often fails to even satisfy the governing equation for all, but the most simple problems. It is thus more of an interesting research topic than a practical tool for applied optimization problems. In the following, we will thus restrict ourselves to the nested formulation.

5.3.1 Non-parametric Optimization

The number of variables does not have to be tied to a fixed set of design features. It can be formulated in a more abstract sense, such that the boundary is allowed to vary by having it defined implicitly as the contour of a spatially varying field, i.e. a *level-set* function. Alternatively, one can use an explicit boundary representation,

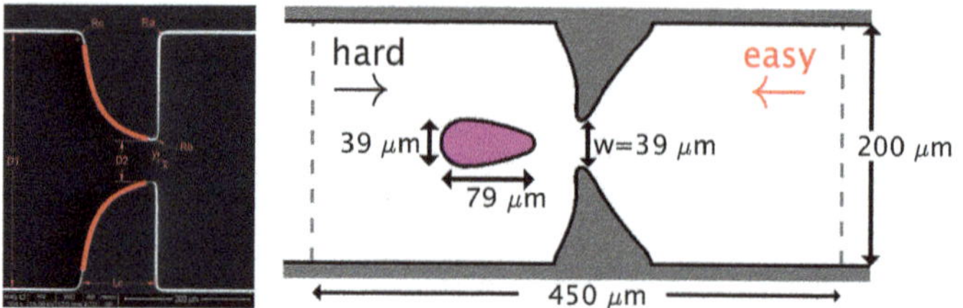

Fig. 5.2 The figure to the *left* shows a typical contraction, which has been extensively studied for its direction dependent hydraulic resistance. The design to the *right* has been studied in the same context, but it features a qualitatively different design, which is the result of topology optimization. The figures are reproduced from [4, 8]

but this normally prevents changes of the design topology. In any case, the number of variables for such an optimization will scale with the numerical discretization of the underlying model, which implies thousands of designs variables and thus also a restricted set of applicable optimization methods. The advantage of these methods is that they allow for extreme design freedom such that the qualitative layout of the design (i.e. the number of holes / the design topology) does not have to be known a priori, hence the name *topology optimization*, see Fig. 5.2.

Examples of this include micro reactors [11] as well as inertial- and viscoelastic rectifiers [8, 12]. If one finds a design with a novel topology using non-parameteric optimization, it is a good idea to perform a parametric optimization as a post-processing step. This can simplify fabrication and enable other researchers to reproduce the experiments. Finally, it is always advisable to

understand the working mechanism of the optimal design and reconsider your problem statement with this in mind.

5.4 Objective Function and Constraints

The choice of objective function, constraints and problem formulation are intimately related. In example, many devices will benefit from the energy that is put into them, so if one imposes fixed in-flow boundary conditions, there is a possibility that the optimization will result in a design with extremely high hydraulic resistance, effectively blocking the system. One can get around this by imposing a pressure drop constraint, or simply by switching to boundary conditions with a fixed pressure drop.

Micro devices for complex fluids tend to rely on effects that become (relatively) stronger at small length scales. This, however, also means that an optimization might try to introduce small length scales and one thus have to consider the choice of design variables and their constraints carefully. In a non-parametric optimization, one might have to introduce a minimum curvature or minimum length scale. In any case, one has to prevent the optimizer from making structures smaller than what can be both accurately captured by the model and experimentally realized.

Finding the best problem statement might involve some trial and error, but

> when you identify potential objective functions and constraints of your problem, you should keep in mind that trivial designs are never optimal for well-posed problem formulations.

I.e. if you want to minimize the viscous dissipation in a channel, the optimal design will either be a complete open or a completely blocked channel depending on whether the flow rate or the pressure drop is fixed, so unless you introduce a volume constraint, you will get a trivial design.

For inequality constraint functions the convention is that they should be negative and the set of design variables respecting all constraints is called the feasible set. Some optimization algorithms satisfy inequality constraints, g_i, by minimizing a modified objective function

$$O' = O + \sum_i^N w_i g_i,$$

where w_i are weights, which are kept as small possible and only increased, when a constraint is violated. The details of such a procedures often involve some heuristics and assumptions about functions and variables. This means that such

> general purpose mathematical optimizers work best, if the problem is stated in a non-dimensional formulation.

Furthermore, equality constraints are often treated as two inequality constraints in the numerical implementation and they always make the problem stiff, so that the maximum step size of the optimizer is severely reduced.

It is a good idea to investigate the smoothness of the problem and whether the objective function is convex or not. If the problem has several minima, as illustrated in Fig. 5.3, it is non-convex and if there are many local minima, it will be difficult to find a good design. Within non-parametric optimization it is common to solve a convex problem, which is similar, but not identical, to the actual problem one wants to solve. One can then make a continuation from the easy and wrong problem to the hard and correct, so that a good design can be found. This tip, however, comes with a footnote:

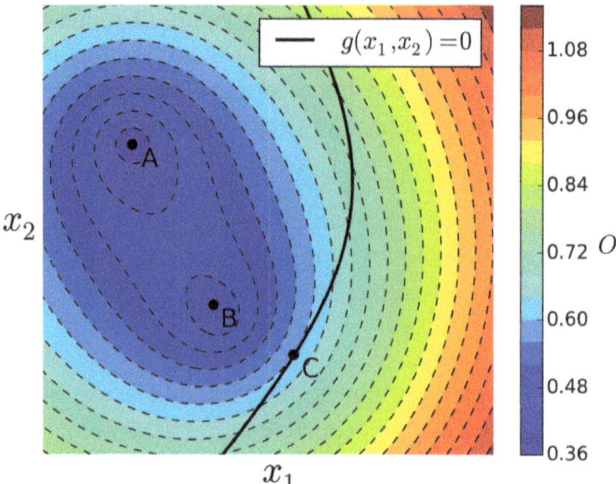

Fig. 5.3 An objective function O is drawn in a contour *plot* as a function of two design variables, x_1 and x_2. The $g(x_1, x_2) = 0$ contour of the constraint function is also sketched and depending on whether the feasible region ($g < 0$) is on the *left* or *right* side of the *plot*, the optimal design is either A or C, respectively. B is a local extremum, so this is an example of a non-convex optimization problem

Local minima are best avoided using continuation methods and multiple initial designs guesses. They, however, tend to be more robust towards parameter- or design variations.

If one wants to take robustness into account in a more systematic way, a range of physical parameters can be considered or different perturbations to the *blueprint design* can be investigated. If several such models are used to construct an objective function, the robustness of the global minimum is likely to improve. The actual value, however, will increase—lunch is never free.

5.4.1 *Pareto-Optimality*

Sometimes there are several critical objective functions, and it is impossible to choose how to prioritise them, before the optimization has been carried out. This is because one wants to know how much can be gained of one objective by giving up a certain amount of another objective, i.e. you accept that *there is no free lunch*, but you want to know the cost. In such a case one will typically resort to multi objective optimization. This involves an objective function based on a weighted average and a detailed study of the effects of the weights. Plotting the objectives as functions of each other will then reveal the pareto optimal front as shown in Fig. 5.4. This also indicates whether some of the optimizations resulted in local minima.

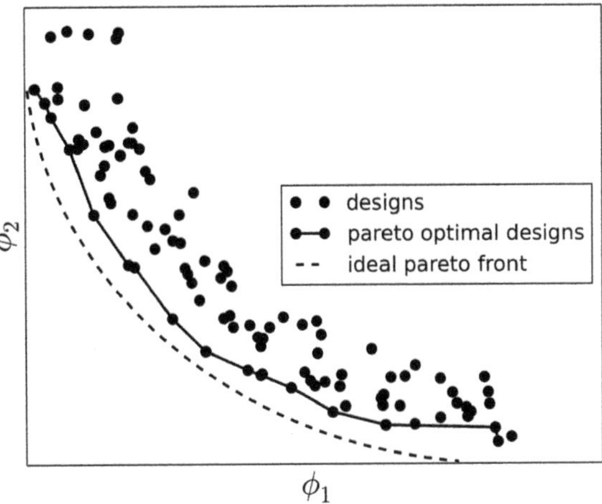

Fig. 5.4 In the context of multiobjective optimization, a design is said to be pareto optimal, when no other design is better in terms of all objectives. The pareto front consists of such designs as illustrated for a problem with two objective functions, ϕ_1 and ϕ_2

5.5 Gradient Free Methods

The most simple optimization one can imagine involves mapping out the entire design space. This is a robust approach that is easy to implement, but it is also very expensive in terms of computational resources. This downside can be somewhat mitigated by using an initial coarse map to select a subregion for further analysis in what can end up as a hierarchical method, but ultimately such an approach is unlikely to be attractive, if one has more than a handful of variables.

Powell's method is a simple gradient free method, which works by

#1 Pick a starting guess $\mathbf{x}_0$ and a number of search vectors $\mathbf{v}_i$.

#2 Perform a line search along each search vector, i.e. find the minimum of $O(w_i\mathbf{v}_i + \mathbf{x}_0)$ with respect to w_i.

#3 Update the guess to $\mathbf{x}_0 + \sum w_i\mathbf{v}_i$ and replace the search vector having the lowest $|w_i|$ with $\sum_i w_i\mathbf{v}_i$. Go to #2.

The search vector substitution is critical for the performance in problems with strong anisotropy, such as the Rosenbrock function, see Fig. 5.5. The complexity of the method lies in the line search, which is also the only part of the algorithm involving function evaluations. The line searches can be performed independently of each other, which makes it trivial to realize a parallel implementation of the algorithm.

The downhill simplex method or Nelder-Mead method is an alternative, which is difficult to realise in a parallel implementation. The concept is also somewhat more complicated, but there is no linear search tolerance to be set and perhaps updated, which is probably why it is the most popular gradient free method. It works using a simplex that is reflected, expanded, contracted and shrunk so as to find the minimum of the objective function.[1] The only parameters of the method are related to the

---

[1] Animations available at https://en.wikipedia.org/wiki/Nelder%E2%80%93Mead_method. Cited 30 April 2017.

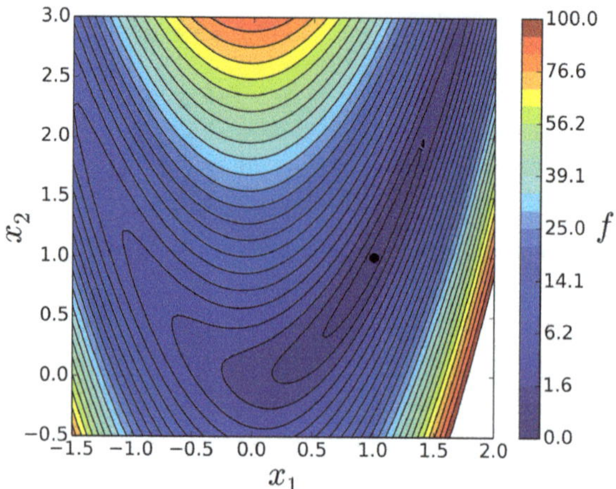

Fig. 5.5 The Rosenbrock function $f(x_1, x_2) = (1 - x_1)^2 + 10(x_2 - x1^2)^2$ has a minimum at $x_1 = x_2 = 1$, but the anisotropic nature of the function makes it difficult to locate this minium, and therefore the function it is a popular benchmark problem for optimization methods

stopping criteria and it is one of the method without internal parameters. Internal parameters can be problematic for poorly scaled problems, so the simplex method is thus an exception to the rule that general purpose mathematical optimizers work best, if the problem is stated in a non-dimensional formulation.

Both the Powell's and the Nelder-Mead method is available through python's `scipy.optimize.minimize` function. Any constraints will have to be enforced by setting the objective function to infinity outside the feasible region.

It is important to test whether different starting guesses results in different minima. The global minimum of the numerical objective function will also change whenever the topology of the discretization is changed, but this effect is reduced with a finer discretization as shown to the left in Fig. 5.6. For a mesh based model this means that the objective function is only smooth, if the connectivity is fixed, but the vertices are allowed to move. Such a strategy has been used to optimize a viscoelastic rectifier [2]. If a minimum is far from the starting guess, the discretization will have to morph a lot, which can be difficult to achieve and reduce the accuracy of the model. In that situation, one should consider making a new discretization and restarting the optimization. To ease this process you should

consider accelerating your parametric studies by using software featuring automatic spatial discretization/meshing.

When using a gradient free method, one can in principle do this in every iteration as illustrated to the right in Fig. 5.6.

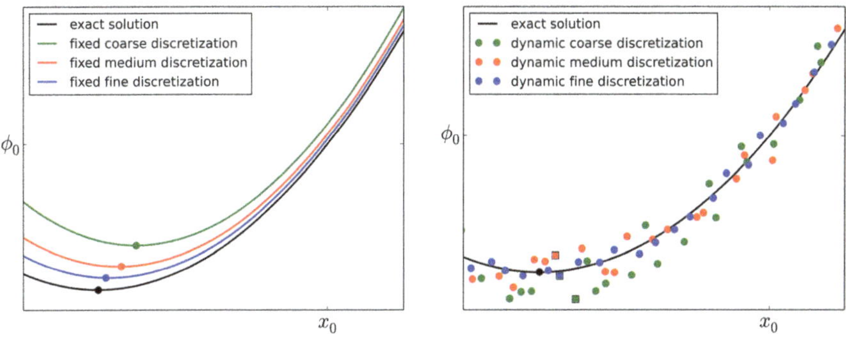

Fig. 5.6 An objective function ϕ is plotted as a function of a variable x. To the *left* a discretization optimized for $x = x_0$ is used, which leads to a low deviation from the exact solution at this point. To the *right* the deviation is low everywhere, because the discretization is changed for every x, but this means that one cannot estimate the gradient accurately using finite differences

5.6 Gradient Based Methods

Gradient based methods require that you can construct a smooth representation of the objective function (at least locally), such that

$$\phi(\mathbf{x}) = \phi(\mathbf{x}_0) + (\mathbf{x} - \mathbf{x}_0) \cdot \nabla\phi|_{\mathbf{x}=\mathbf{x}_0} + \mathcal{O}\left((\mathbf{x} - \mathbf{x}_0)^2\right)$$

i.e. a 1st order Taylor expansion. The most simple way to calculate the components of the gradient is using finite differences, i.e.

$$
\begin{aligned}
\left(\nabla\phi|_{\mathbf{x}=\mathbf{x}_0}\right)_i &= \frac{\phi(\mathbf{x}_0 + \delta\mathbf{x}_i) - \phi(\mathbf{x}_0)}{\delta x} + \mathcal{O}(\delta x), && \text{forward difference} \\
&= \frac{\phi(\mathbf{x}_0 + \delta\mathbf{x}_i) - \phi(\mathbf{x}_0 - \delta\mathbf{x}_i)}{2\delta x} + \mathcal{O}(\delta x)^2, && \text{central difference,}
\end{aligned}
$$

where $\delta\mathbf{x}_i$ is a small variation in the i'th variable.[2] One can use this technique to test the accuracy of sensitivities, if they are computed without using finite differences.

The most simple approach to gradient based optimization is gradient descent, which boils down to a line search along the gradient direction. More interesting is the conjugate gradient (CG) method, which is guaranteed to converge in the same

[2]A value three orders of magnitude larger than the machine precision is a good starting point, but the optimal value is problem dependent and it is thus a good idea to study, when numerical noise dies out and 2nd order effects sets in.

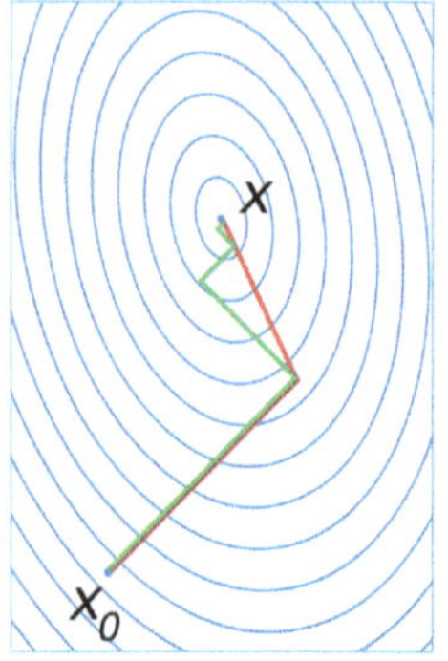

Fig. 5.7 A gradient descent algorithm is compared to the conjugate gradient method for a quadratic function in 2D. It will thus converge in 2 iterations, once it is close to a minimum. https://en.wikipedia.org/wiki/Conjugate_gradient_method. Cited 30 April 2017

number of iterations as the number of design variables, if the objective function is quadratic (which is generally the case close to the minimum), see Fig. 5.7. The CG method is also available through python's `scipy.optimize.minimize` function.

If one can also calculate the Hessian, the 2nd order Taylor expansion can be constructed. It is straightforward to compute the minimum of this quadratic function and iterating in this way is called Newton's method. This has quadratic order asymptotic convergence, which means that the logarithm of the error is halved in every iteration, when the method is close to an optimum. In comparison, gradient descent will only halve the actual error in every iteration.

For most applications, the convergence rate of a method is irrelevant, since this only applies close to the minimum, while most of the computational time tends to be spent far from the minimum. If faced with the question of which optimization algorithm to pick for some arbitrary problem, it is thus relevant to look at the performance for a wide range of benchmark problems as those found in the CUTEr (Constrained and Unconstrained Testing Environment, revisited) set, which contains around 1,000 optimization problems. The Ipopt package does well on this test set [3], it is open source and available through the PyIpopt python module.[3] The method of moving asymptotes is an alternative that is very popular within structural optimization [13], but it is only free for academic use.

5.6.1 Non-parametric Optimization

Non-parametric optimization methods tend to involve many design variables and therefore one has to use efficient techniques for the computation of the gradient. The adjoint variable technique is the only method for doing this, either in a discrete or a continuous version. Both can give the exact gradient for a given discretization, but it

---

[3]https://github.com/xuy/pyipopt. Cited 30 April 2017.

is only guaranteed for the discrete versions. Dolfin adjoint is a module for the FEniCS [1] finite element package, and it can calculate the discrete gradient automatically [6]. In general,

> Software capable of automatic differentiation is a great help for non-linear problems as well as for computation of gradients.

The advantage of the continuous adjoint technique is that it is independent of numerical methods and thus potentially provides insight on a higher level. As an example, we will consider Stokes flow with a Brinkman damping term. The sensitivity of the viscous dissipation with respect to local changes in damping coefficient can be calculated using the continuous adjoint technique as follows. The governing equations and objective function, ϕ are

$$\phi = \int_\Omega \left(\tfrac{1}{2}\underline{\underline{\varepsilon}} : \underline{\underline{\sigma}} - \mathbf{u} \cdot \mathbf{F} \right) d\Omega \tag{5.1}$$

$$0 = \nabla \cdot \underline{\underline{\sigma}} + \mathbf{F} \tag{5.2}$$

$$0 = \nabla \cdot \mathbf{u} \quad \text{where} \tag{5.3}$$

$$\underline{\underline{\sigma}} = -\underline{\underline{I}}p + \eta\underline{\underline{\varepsilon}}, \quad \underline{\underline{\varepsilon}} = \nabla\mathbf{u} + [\nabla\mathbf{u}]^T$$

$$\mathbf{F} = -\alpha\mathbf{u},$$

where $\mathbf{u}$, p, η, $\mathbf{F}$, $\underline{\underline{\sigma}}$, $\underline{\underline{\varepsilon}}$ and α are the velocity, pressure, viscosity, volumetric force, stress, rate of deformation and damping term, respectively. It is easy to see that the objective function can be simplified due to Eq. (5.3),

$$\phi = \int_\Omega \left(\tfrac{1}{2}\eta\underline{\underline{\varepsilon}} : \underline{\underline{\varepsilon}} - \mathbf{u} \cdot \mathbf{F} \right) d\Omega, \quad \text{because} \quad \underline{\underline{\varepsilon}} : \underline{\underline{I}} = 2\nabla \cdot \mathbf{u} = 0 \tag{5.4}$$

A variation in the damping term, $\delta\alpha$ will result in a variation of the objective function,

$$\delta\phi = \int_\Omega \left(\eta\frac{\partial\underline{\underline{\varepsilon}}}{\partial\alpha} : \underline{\underline{\varepsilon}} - \frac{\partial\mathbf{u}}{\partial\alpha} \cdot \mathbf{F} - \mathbf{u} \cdot \frac{\partial\mathbf{F}}{\partial\alpha} \right) \delta\alpha\, d\Omega$$

$$= \int_\Omega \left(\eta\frac{\partial\underline{\underline{\varepsilon}}}{\partial\alpha} : \underline{\underline{\varepsilon}} + 2\alpha\mathbf{u} \cdot \frac{\partial\mathbf{u}}{\partial\alpha} + \mathbf{u}^2 \right) \delta\alpha\, d\Omega \tag{5.5}$$

This can be expanded further, but the point is that the derivatives of $\mathbf{u}$ and p with respect to α are unknown and therefore the derivative of $\underline{\underline{\varepsilon}}$ is also unknown, so they somehow have to be eliminated. This can be achieved by constructing other partial differential equations with the same terms and adding/subtraction the equations. The starting point of this procedure is the introduction the adjoint variables $\tilde{\mathbf{u}}$ and $\tilde{p}$, which by definition are invariant with respect to the $\delta\alpha$ variation. Multiplying $\tilde{\mathbf{u}}$ with Eq. (5.2) and integrating over the domain, Ω, yields

$$
\begin{aligned}
0 &= \int_\Omega \tilde{\mathbf{u}} \cdot \left(\nabla \cdot \underline{\underline{\sigma}} + \mathbf{F} \right) d\Omega \\
&= \int_{\partial\Omega} \tilde{\mathbf{u}} \cdot \underline{\underline{\sigma}} \cdot \hat{\mathbf{n}} ds - \int_\Omega \left(\nabla\tilde{\mathbf{u}} : \underline{\underline{\sigma}} + \tilde{\mathbf{u}} \cdot \mathbf{F} \right) d\Omega \\
&= \int_{\partial\Omega} \tilde{\mathbf{u}} \cdot \underline{\underline{\sigma}} \cdot \hat{\mathbf{n}} ds - \int_\Omega \left(\tfrac{1}{2}\underline{\underline{\tilde{\varepsilon}}} : \underline{\underline{\sigma}} - \tilde{\mathbf{u}} \cdot \mathbf{F} \right) d\Omega \quad \text{where} \quad \underline{\underline{\tilde{\varepsilon}}} = \nabla\tilde{\mathbf{u}} + \left[\nabla\tilde{\mathbf{u}} \right]^T \quad (5.6)
\end{aligned}
$$

where we have used the divergence theorem and the fact that the stress tensor is symmetric. Taking the variation with respect to $\delta\alpha$ yields

$$
\begin{aligned}
0 = &\int_{\partial\Omega} \tilde{\mathbf{u}} \cdot \frac{\partial \underline{\underline{\sigma}}}{\partial \alpha} \cdot \hat{\mathbf{n}}\delta\alpha ds - \int_\Omega \left(\tfrac{1}{2}\underline{\underline{\tilde{\varepsilon}}} : \frac{\partial \underline{\underline{\sigma}}}{\partial \alpha} - \tilde{\mathbf{u}} \cdot \frac{\partial \mathbf{F}}{\partial \alpha} \right) \delta\alpha d\Omega \\
= &\int_{\partial\Omega} \tilde{\mathbf{u}} \cdot \frac{\partial \underline{\underline{\sigma}}}{\partial \alpha} \cdot \hat{\mathbf{n}}\delta\alpha ds \\
&- \int_\Omega \left(\tfrac{1}{2}\underline{\underline{\tilde{\varepsilon}}} : \left[-\underline{\underline{\mathbf{I}}}\frac{\partial p}{\partial \alpha} + \eta\frac{\partial \underline{\underline{\varepsilon}}}{\partial \alpha} \right] + \tilde{\mathbf{u}} \cdot \left[\alpha\frac{\partial \mathbf{u}}{\partial \alpha} + \mathbf{u} \right] \right) \delta\alpha d\Omega \quad (5.7)
\end{aligned}
$$

It is now easy to see that we can eliminate all derivatives with respect to α by adding Eqs. (5.5) and (5.7) and assuming $\tilde{\mathbf{u}} = 2\mathbf{u}$. This means that $\underline{\underline{\tilde{\varepsilon}}} : \underline{\underline{\mathbf{I}}} = 2\nabla \cdot \tilde{u} = 0$, so we get

$$
\delta\phi = \int_{\partial\Omega} \tilde{\mathbf{u}} \cdot \frac{\partial \underline{\underline{\sigma}}}{\partial \alpha} \cdot \hat{\mathbf{n}}\delta\alpha ds + \int_\Omega -\mathbf{u}^2 \delta\alpha d\Omega
$$

The boundary term drops out, if we restrict ourselves to no-slip boundary conditions and inlet/outlets with fixed pressure and zero normal viscous stress. This is due to the fact that either $\tilde{\mathbf{u}} = 2\mathbf{u} = \mathbf{0}$ or $\underline{\underline{\sigma}} \cdot \hat{\mathbf{n}} = \hat{\mathbf{n}}p_{\text{bnd}}$. The sensitivity of the objective function thus becomes

$$
\frac{\partial\phi}{\partial\alpha} = -\mathbf{u}^2
$$

The fact that the adjoint velocity can be expressed explicitly from the physical velocity means that the problem is *self-adjoint*, which is a special case. In general the sensitivity will depend on the physical as well as the adjoint variables with separate partial differential equations and associated boundary conditions for the adjoint variables. The adjoint equations are, however, guaranteed to be linear. For an example of this refer to the Appendix A.

5.7 Summary

In this chapter we have covered when optimization makes sense and how it is best carried out. The following key tips have been given

#1 Verify the robustness of your numerical model and never perform optimization outside its limits—the optimization result should be independent of the numerical discretization.

#2 You are unlikely to benefit from an optimization, if you have an inaccurate model and little design freedom.

#3 Try to select the design variables of your problem such that box constraints can be used.

#4 Understand the working mechanism of the optimal design and reconsider your problem statement with this in mind.

#5 When you identify potential objective functions and constraints of your problem, you should keep in mind that trivial designs are never optimal for well-posed problem formulations.

#6 Local minima are best avoided using continuation methods and multiple initial designs guesses. They, however, tend to be more robust towards parameter- or design variations.

#7 General purpose mathematical optimizers work best, if the problem is stated in a non-dimensional formulation.

#8 Consider accelerating your parametric studies by using software featuring automatic spatial discretization/meshing.

#9 Test the accuracy of sensitivities, if they are computed without using finite differences.

#10 Software capable of automatic differentiation is a great help for non-linear problems as well as for computation of gradients.

Other points relate to multi objective optimization, the nested optimization formulation and the adjoint technique for computing sensitivities.

5.8 Exercises

Discussion exercises
Answer the following questions with regards to a project

#1 Do you have a model? If not, do you know enough about your system to construct a model? How accurate and robust is or could it be? Do you have a lot of design freedom?

#2 Discuss possible objective functions, variables and constraints for your project.

#3 Do you understand the working mechanism of your system? Can you predict the outcome of an optimization in a qualitative sense?

Numerical exercises
Perform the following exercises in your software of choice

#1 Implement Powel's method and test it for the Rosenbrock function. Plot the path for different initial guesses.

#2 Reuse one of your models from the previous chapter and choose an objective function and a constraint. Plot the relation between the two with a fixed and dynamic mesh topology.

#3 Use an optimization method to find the minimum automatically.

Acknowledgements This work is supported by the Villum Foundation (Grant No. 9301) and the Danish Council for Independent Research (DNRF122).

References

1. Alnæs, M., Blechta, J., Hake, J., Johansson, A., Kehlet, B., Logg, A., et al. (2015). The FEniCS project version 1.5. *Archive of Numerical Software*, *3*(100): 9–23.
2. Alves, M.A. (2011). Design of optimized microfluidic devices for viscoelastic fluid flow. In *Technical Proceedings of the 2011 NSTI Nanotechnology Conference and Expo. 2*: 474–477.
3. Bongartz, I., Conn, A. R., Gould, N., & Toint, Ph L. (1995). CUTE: Constrained and unconstrained testing environment. *ACM Transactions on Mathematical Software (TOMS)*, *21*(1), 123–160.
4. Campo-Deaño, L., Galindo-Rosales, F. J., Pinho, F. T., Alves, M. A., & Oliveira, M. S. N. (2011). Flow of low viscosity Boger fluids through a microfluidic hyperbolic contraction. *Journal of Non-Newtonian Fluid Mechanics*, *166*(21), 1286–1296.
5. Evgrafov, A. (2015). On Chebyshev's method for topology optimization of Stokes flows. *Structural and Multidisciplinary Optimization*, *51*(4), 801–811.
6. Farrell, P. E., Ham, D. A., Funke, S. W., & Rognes, M. E. (2013). Automated derivation of the adjoint of high-level transient finite element programs. *SIAM Journal on Scientific Computing*, *35*(4), C369–C393.
7. Galindo-Rosales, F. J., Oliveira, M. S. N., & Alves, M. A. (2014). Optimized cross-slot microdevices for homogeneous extension. *RSC Advances*, *4*(15), 7799–7804.
8. Jensen, K. E., Szabo, P., Okkels, F., & Alves, M. A. (2012). Experimental characterisation of a novel viscoelastic rectifier design. *Biomicrofluidics*, *6*(4), 044112.
9. Jensen, K. E., Szabo, P., & Okkels, F. (2012). Topology optimization of viscoelastic rectifiers. *Applied Physics Letters*, *100*(23), 234102.
10. Jensen, K. E., Szabo, P., & Okkels, F. (2014). Optimization of bistable viscoelastic systems. *Structural and Multidisciplinary Optimization*, *49*(5), 733–742.
11. Krühne, U., Larsson, H., Heintz, S., Ringborg, R. H., Rosinha, I. P., Bodla, V. K., et al. (2014). Systematic development of miniaturized (bio) processes using Process Systems Engineering (PSE) methods and tools. *Chemical and Biochemical Engineering Quarterly*, *28*(2), 203–214.
12. Lin, S., Zhao, L., Guest, J. K., Weihs, T. P., & Liu, Z. (2015). Topology optimization of fixed-geometry fluid diodes. *Journal of Mechanical Design*, *137*(8), 081402.
13. Svanberg, K. (1987). The method of moving asymptotes-A new method for structural optimization. *International Journal for Numerical Methods in Engineering*, *24*(2), 359–373.

Appendix A
Sensitivity for Stokes Flow

Consider the following optimization problem, where the stokes flow in some point, $\mathbf{x}_0$ is minimized along the x-direction.

$$\phi = \int_\Omega \delta(\mathbf{x} - \mathbf{x}_0)\mathbf{u} \cdot \hat{\mathbf{x}}d\Omega \tag{A.1}$$

$$\mathbf{0} = \nabla \cdot \underline{\underline{\sigma}} + \mathbf{F}$$

$$0 = \nabla \cdot \mathbf{u} \quad \text{where} \tag{A.2}$$

$$\underline{\underline{\sigma}} = -\underline{\underline{I}}p + \eta\underline{\underline{\varepsilon}}, \quad \underline{\underline{\varepsilon}} = \nabla\mathbf{u} + [\nabla\mathbf{u}]^T$$

$$\mathbf{F} = -\alpha\mathbf{u},$$

A variation in the damping term, $\delta\alpha$ will result in a variation of the objective function,

$$\delta\phi = \int_\Omega \delta(\mathbf{x} - \mathbf{x}_0)\frac{\partial\mathbf{u}}{\partial\alpha} \cdot \hat{\mathbf{x}}\delta\alpha d\Omega \tag{A.3}$$

We cannot evaluate the derivative of u with respect to α, so it somehow has to be eliminated. This can be achieved by constructing other partial differential equations with the same terms and adding/subtraction the equations. The starting point of this procedure is the introduction the adjoint variables $\tilde{\mathbf{u}}$ and $\tilde{p}$, which by definition are invariant with respect to the $\delta\alpha$ variation. Multiplying $\tilde{\mathbf{u}}$ with Eq. (5.2) and integrating over the domain, Ω, yields

© Springer International Publishing AG 2018

F.J. Galindo-Rosales (ed.), *Complex Fluid-Flows in Microfluidics*,

DOI 10.1007/978-3-319-59593-1

$$
\begin{aligned}
0 &= \int_{\Omega} \tilde{\mathbf{u}} \cdot \left(\nabla \cdot \underline{\underline{\sigma}} + \mathbf{F} \right) d\Omega \\
&= \int_{\partial\Omega} \tilde{\mathbf{u}} \cdot \underline{\underline{\sigma}} \cdot \hat{\mathbf{n}} ds - \int_{\Omega} \left(\nabla\tilde{\mathbf{u}} : \underline{\underline{\sigma}} + \tilde{\mathbf{u}} \cdot \mathbf{F} \right) d\Omega \\
&= \int_{\partial\Omega} \tilde{\mathbf{u}} \cdot \underline{\underline{\sigma}} \cdot \hat{\mathbf{n}} ds - \int_{\Omega} \left(\tfrac{1}{2}\underline{\underline{\tilde{\varepsilon}}} : \underline{\underline{\sigma}} - \tilde{\mathbf{u}} \cdot \mathbf{F} \right) d\Omega \quad \text{where} \quad \underline{\underline{\tilde{\varepsilon}}} = \nabla\tilde{\mathbf{u}} + \left[\nabla\tilde{\mathbf{u}} \right]^T \\
&= \int_{\partial\Omega} \tilde{\mathbf{u}} \cdot \underline{\underline{\sigma}} \cdot \hat{\mathbf{n}} ds - \int_{\Omega} \left(\tfrac{1}{2}\underline{\underline{\tilde{\varepsilon}}} : \left[-\underline{\underline{\mathbf{I}}}p + 2\eta\nabla\mathbf{u} \right] - \tilde{\mathbf{u}} \cdot \mathbf{F} \right) d\Omega \\
&= \int_{\partial\Omega} \left(\tilde{\mathbf{u}} \cdot \underline{\underline{\sigma}} - \eta\mathbf{u} \cdot \underline{\underline{\tilde{\varepsilon}}} \right) \hat{\mathbf{n}} ds - \int_{\Omega} \left(-p\nabla \cdot \tilde{\mathbf{u}} - \eta\mathbf{u} \cdot \nabla \cdot \underline{\underline{\tilde{\varepsilon}}} - \tilde{\mathbf{u}} \cdot \mathbf{F} \right) d\Omega
\end{aligned}
$$

where we have used the divergence theorem and the fact that the stress tensor is symmetric. Taking the variation with respect to $\delta\alpha$ yields

$$
\begin{aligned}
0 &= \int_{\partial\Omega} \left(\tilde{\mathbf{u}} \cdot \frac{\partial\underline{\underline{\sigma}}}{\partial\alpha} - \eta\frac{\partial\mathbf{u}}{\partial\alpha} \cdot \underline{\underline{\tilde{\varepsilon}}} \right) \cdot \hat{\mathbf{n}}\delta\alpha ds + \int_{\Omega} \left(\frac{\partial p}{\partial\alpha}\nabla \cdot \tilde{\mathbf{u}} + \eta\frac{\partial\mathbf{u}}{\partial\alpha} \cdot \nabla \cdot \underline{\underline{\tilde{\varepsilon}}} + \tilde{\mathbf{u}} \cdot \frac{\partial\mathbf{F}}{\partial\alpha} \right) \delta\alpha d\Omega \\
&= \int_{\partial\Omega} \left(\tilde{\mathbf{u}} \cdot \frac{\partial\underline{\underline{\sigma}}}{\partial\alpha} - \eta\frac{\partial\mathbf{u}}{\partial\alpha} \cdot \underline{\underline{\tilde{\varepsilon}}} \right) \cdot \hat{\mathbf{n}}\delta\alpha ds \\
&\quad + \int_{\Omega} \left(\frac{\partial p}{\partial\alpha}\nabla \cdot \tilde{\mathbf{u}} + \eta\frac{\partial\mathbf{u}}{\partial\alpha} \cdot \nabla \cdot \underline{\underline{\tilde{\varepsilon}}} + \tilde{\mathbf{u}} \cdot \left[-\mathbf{u} - \alpha\frac{\partial\mathbf{u}}{\partial\alpha} \right] \right) \delta\alpha d\Omega
\end{aligned}
\tag{A.4}
$$

Repeating the process for Eq. (A.2) gives

$$
\begin{aligned}
0 &= \int_{\Omega} \tilde{p}\nabla \cdot \mathbf{u} d\Omega \\
&= \int_{\partial\Omega} \tilde{p}\mathbf{u} \cdot \hat{\mathbf{n}} ds - \int_{\Omega} \mathbf{u} \cdot \nabla\tilde{p} d\Omega \\
&= \int_{\partial\Omega} \tilde{p}\frac{\partial\mathbf{u}}{\partial\alpha} \cdot \hat{\mathbf{n}}\delta\alpha ds - \int_{\Omega} \frac{\partial\mathbf{u}}{\partial\alpha} \cdot \nabla\tilde{p}\delta\alpha d\Omega
\end{aligned}
\tag{A.5}
$$

Adding Eqs. (A.3), (A.4) and (A.5) gives

$$
\begin{aligned}
\delta\phi &= \int_{\Omega} \left(\frac{\partial\mathbf{u}}{\partial\alpha} \cdot \left[-\nabla\tilde{p} + \eta\nabla \cdot \underline{\underline{\tilde{\varepsilon}}} + \delta(\mathbf{x} - \mathbf{x}_0)\hat{\mathbf{x}} - \alpha\tilde{\mathbf{u}} \right] - \tilde{\mathbf{u}} \cdot \mathbf{u} + \frac{\partial p}{\partial\alpha}\nabla \cdot \tilde{\mathbf{u}} \right) \delta\alpha d\Omega \\
&\quad + \int_{\partial\Omega} \left(\tilde{\mathbf{u}} \cdot \frac{\partial\underline{\underline{\sigma}}}{\partial\alpha} + \frac{\partial\mathbf{u}}{\partial\alpha} \cdot \left(\overbrace{\underline{\underline{\mathbf{I}}}\tilde{p} - \eta\underline{\underline{\tilde{\varepsilon}}}}^{-\underline{\underline{\tilde{\sigma}}}} \right) \right) \cdot \hat{\mathbf{n}}\delta\alpha ds,
\end{aligned}
\tag{A.6}
$$

so the sensitivity becomes

$$\frac{\partial \phi}{\partial \alpha} = -\tilde{\mathbf{u}} \cdot \mathbf{u} \quad \text{where} \quad \mathbf{0} = \nabla \cdot \tilde{\underline{\underline{\sigma}}} + \tilde{\mathbf{F}} \quad \text{and} \quad 0 = \nabla \cdot \tilde{\mathbf{u}},$$

$$\tilde{\underline{\underline{\sigma}}} = -\underline{\underline{I}}\tilde{p} + \eta \tilde{\underline{\underline{\varepsilon}}}, \quad \tilde{\mathbf{F}} = \delta(\mathbf{x} - \mathbf{x}_0)\hat{\mathbf{x}} - \alpha \tilde{\mathbf{u}}$$

$$0 = \left(\tilde{\mathbf{u}} \cdot \frac{\partial \underline{\underline{\sigma}}}{\partial \alpha} - \eta \frac{\partial \mathbf{u}}{\partial \alpha} \cdot \tilde{\underline{\underline{\sigma}}} \right) \cdot \hat{\mathbf{n}} \quad \text{on} \quad \partial \Omega \quad \text{i.e.}$$

$$\tilde{u} = \mathbf{0}, \quad \text{on} \quad \partial \Omega_{\mathbf{u} = \mathbf{u}_{\mathrm{bnd}}} \quad \text{and} \quad \tilde{\underline{\underline{\sigma}}} \cdot \hat{\mathbf{n}} = \mathbf{0} \quad \text{on} \quad \partial \Omega_{\underline{\underline{\sigma}} = -\underline{\underline{I}}p_{\mathrm{bnd}}}.$$

In other words the adjoint flow is driven solely by the point force at $\mathbf{x}_0$. This drives the flow forward. The sensitivity is independent of the adjoint pressure, but in practice we need some constraint on it to solve the system. It could be fixed at some arbitrary point or its integral equal to zero could be added as an additional constraint in the linearised system.

www.ingramcontent.com/pod-product-compliance
Ingram Content Group UK Ltd.
Pitfield, Milton Keynes, MK11 3LW, UK
UKHW021458010426
11640UKWH00022BB/374